One of the central questions of physics is whether or not a Theory of Everything is possible. Many physicists believe that such a theory might be attainable, a belief which has lead to speculation that we might one day 'know the mind of God'. But what would be the philosophical implications of having a blueprint for the Universe? What does physics tell us about reality? Does possession of the Theory of Everything leave room for the existence of God? In this fascinating book, a group of distinguished physicists and philosophers examine not only the claims of modern physics, but also the impact these claims have on our view of the world.

Based on talks given at the Third Erasmus Ascension Symposium in the Netherlands, the book contains contributions from John Barrow, Paul Davies, Dennis Dieks, Willem Drees, Paul Feyerabend, Bas van Fraassen, Mary Hesse, Gerard 't Hooft and Ernan McMullin. Also included, however, are the discussions which followed the talks, characterized by a frank exchange of views and many clear insights into these difficult issues.

At a time when many people view science with deep suspicion, this book will be of great interest to anyone wishing to explore the complex relationships that exist between physics and philosophy, theology and ideology.

# Physics and our view of the world

# Physics and our view of the world

edited by Jan Hilgevoord

*Emeritus Professor of the Foundations of Physics*
*University of Utrecht*

Published by the Press Syndicate of the University of Cambridge
The Pitt Building, Trumpington Street, Cambridge CB2 1RP
40 West 20th Street, New York, NY 10011–4211, USA
10 Stamford Road, Oakleigh, Melbourne 3166, Australia



First published 1994

Printed in Great Britain at the University Press, Cambridge

*A catalogue record for this book is available from the British Library*

*Library of Congress cataloguing in publication data*

Physics and our view of the world / edited by Jan Hilgevoord.
p. cm.
'This book is a result of the third Erasmus Ascension Symposium organized by the Praemium Erasmianum Foundation in Amsterdam' – Foreword.
Includes index.
ISBN 0 521 45372 0
1. Physics – Philosophy – Congresses. 2. Science – Philosophy – Congresses.
3. Religion and science – Congresses. I. Hilgevoord, Jan. II. Praemium Erasmianum Foundation.
QC5.56.P486 1995
530′. 01–dc20 93–46713 CIP

ISBN 0 521 45372 0 hardback
ISBN 0 521 47680 1 paperback

WV

# Contents

# Foreword

This book is the result of the third Erasmus Ascension Symposium organized by the Praemium Erasmianum Foundation in Amsterdam. The Praemium Erasmianum is known primarily for the Erasmus Prize which, since 1958, has been awarded annually to individuals or organizations in Europe that have exceptionally distinguished themselves in the field of European culture. In addition to this activity, every two years the Foundation also organizes a symposium in the Netherlands concerning a specially chosen multi-disciplinary topic of current interest. In 1992, in conjunction with a day that was open to the public, a select group of 40 young Dutch scholars once again had the opportunity to engage in a penetrating and in-depth discussion on the chosen topic with a number of specialists during the three-day symposium.

The choice of the theme for the 1992 symposium, Physics and Our View of the World, was prompted by the great interest that exists for this topic, as is evident from the appearance of so many semi-popular publications in this field. In this way, the Foundation hoped to provide a forum for the useful exchange of views and ideas about the philosophical and religious implications of recent developments in modern physics. We thus hoped to help bridge the gap between the realms of the exact and the spiritual sciences, a division which is still so apparent in our time.

We are pleased that Professor Jan Hilgevoord, faculty member of the Department of the History and Foundations of Mathematics and Science at the University of Utrecht, was able to organize this symposium for the Foundation.

The publication of this book, will, we hope, make the valuable contributions to this symposium accessible to a much wider public.

H. R. Hoetink

*Director, Praemium Erasmianum Foundation*

# Introduction

Jan Hilgevoord

Until quite recently physicists rarely speculated about relations between the laws of fundamental physics, the physics of elementary particles and the Universe, and our existence on earth as conscious beings. They admitted that no physical reason for our existence was known, and that man was an alien in the physical world, but they did not perceive any conflict between our existence and the basic laws of physics.

A first attempt to connect humankind and elementary particle physics was Fritjof Capra's *The Tao of Physics* (1975). In this book Capra tried to relate the so-called 'bootstrap' theory of elementary particles to Eastern mysticism by pointing to similarities between the picture that bootstrap physics gives of the behaviour of elementary particles and the utterances of certain Eastern mystics. To Capra these similarities suggested a deep analogy between the world at the level of elementary particles and the world of personal experience. Such a relation would lessen the deplorable alienation of advanced abstract physics from what can be directly and personally experienced.

*The Tao of Physics* was a great success. The book was welcomed by supporters of holism and of the New Age Movement. It has been translated into many languages and has been followed by related books by Capra and others, but it was conspicuously ignored by the physicists, who found it at best wishful thinking and a lot of nonsense and at worst pure deception. But the book caused a marked increase in public interest in the results of modern physics. This interest, in turn, has led physicists who are specialists in elementary particle physics and cosmology to try and explain these results to the general public; in so doing, some also gave their views on what this remote world of strange particles and abstract theory might imply for our perspective of the world and ourselves. Probably best known among this second wave of publications is the series of books by Paul Davies bearing titles like *God and the New Physics* and *The Mind of God*, and, of course, Stephen Hawking's all-time bestseller *A Brief History of Time*. In this book Hawking seriously considers the possibility of a unifying physical theory, a Theory of Everything, and in the final paragraph he writes that when we have such a theory: 'Then we shall all, philosophers, scientists, and just ordinary people, be able to take part in the discussion of the question of why it is that we and the Universe exist. If we find the answer to that, it would be the ultimate triumph of human reason – for then we would know the mind of God.'

Whereas Capra, at the time when he wrote *The Tao of Physics*, was a relatively unknown physicist, Hawking is a leading figure in cosmology, and whatever he says is of interest to the physics community. Nevertheless, I don't think that the majority of scientists take seriously his statements relating God to the final physical theory. Some even felt compelled to protest against such arrogance and to testify to their belief in a less abstract God. Hawking's view on God may not be representative of the attitudes of physicists in general, but his belief that a Theory of Everything may be attainable is shared by many experts in the field. It is hard, then, to avoid thinking about the consequences such a theory would have for our view of the world and our own role in it as creatures who were capable of discovering this ultimate truth about the world. It was the purpose of the Third Erasmus Ascension Symposium to provide first-hand information regarding the status of such theories and their philosophical implications. To that end a number of prominent physicists and philosophers of science were invited to discuss these problems with each other and with the participants of the symposium.

It is rare for physicists and philosophers to engage in direct discussions; they speak different languages. But the philosophers at this symposium all had a very sound knowledge of physics and some of them had even started their careers as physicists before they turned to the philosophy of science. So, on this occasion there was no language problem. To me this seemed a tremendous advantage. Many scientists, especially physicists, do not have a high opinion of philosophy. To them the fantastic amount of highly non-trivial knowledge that scientists have acquired over the last few hundred years bears no comparison with the seemingly endless squabbles about ever more subtle points of no practical importance that philosophers have spent their time on. Of course, this is being unfair to philosophers. Philosophical questions cannot be answered by empirical inquiry and the pace of progress in philosophy, therefore, is much slower than in physics. But anyone interested in what physics tells us about the world can hardly avoid asking philosophical questions. Many great physicists have occupied themselves with philosophy; the names of Einstein, Bohr, Schrödinger and Heisenberg come to mind. Einstein said that physics without philosophy was not interesting enough to spend a lifetime on.

The main question in the philosophy of physics is: 'what do we know when we know physics?' This is the question to ask when confronted with the remarkable recent claims of physicists and it was also the basic question of the symposium. It will become evident from the contributions in this book that questions like this are at least as hard to answer as any questions in physics.

The symposium was held on 27–31 May 1992, in Leiden and Oosterbeek in the Netherlands. There were three main themes:

(1) What do the results of modern physics tell us about reality?
(2) Does the scientific view of the world have a special status compared to other views?
(3) Physics and theology.

The theoretical physicist Gerard 't Hooft, the astrophysicist John Barrow and the philosopher Ernan McMullin were asked to talk on the first theme. The philosophers Paul Feyerabend and Bas van Fraassen were invited to speak on the second theme, and the theoretical physicist Paul Davies and the philosopher Mary Hesse agreed to talk on the third theme. The physicists were asked to give particular attention to the idea of a Theory of Everything, but otherwise the speakers were free to develop their contributions as they wished. There was no preliminary consultation. The result was a very interesting collection of lectures; clearly, the subject stimulated the speakers to give rather personal views. To enhance the coherence of the symposium, an introduction to the second theme was given by Dennis Dieks who is a philosopher of science, and to the third theme by Willem Drees who is a philosopher of religion.

The discussion was an essential part of the symposium and helped considerably to clarify the lectures and to obtain specific answers from the speakers. A selection of questions and replies from the discussion is included in this book.

In the remainder of this introduction I shall say a few words about the contents of each of the lectures. I shall also try and connect them. But my main purpose, of course, is to stimulate the reader to read the book.

The contributions in this book are ordered according to the three themes mentioned above, but there are many overlaps. Gerard 't Hooft begins the series with an admirably clear and thought-provoking discussion of Theories of Everything: what they are, what they can do and what they can't. When writing for a general public on a subject in theoretical physics, physicists are greatly handicapped: they cannot speak their own mathematical language. They are thus forced to think up pictures and analogies that convey the characteristics of their physical theories. 't Hooft uses Conway's Game of Life, a computer game, as a model of a Theory of Everything. The game shows how simple rules may lead to surprisingly complex outcomes. 't Hooft believes that a Theory of Everything is possible in principle, although we may never be able to find it or make practical use of it. If such a theory exists it would mean that Nature is like a very big and fast computer. But a theory of this type will not be a Theory of Everything in the literal sense: it will not enable us to explain just anything that we should want to explain. Because the reader should not be left with the impression that the rules of present-day physics are as simple as those of Conway's game, 't Hooft briefly discusses the Standard Model of elementary particles, as well as its shortcomings as a Theory of Everything. Of the many things that are touched upon in this rich article I mention the author's dissenting view on quantum mechanics and his astounding speculations regarding the nature of time and the merging of theoretical physics with pure mathematics.

In the second contribution the astrophysicist John Barrow looks at Theories of Everything from a cosmologist's perspective. Like 't Hooft, he stresses the limitations that are inherent in the elementary particle physicists' concept of a Theory of Everything. He also points to the formidable problems one is likely to encounter in searching for such a theory. It may well be, for example, that the part of the Universe that we are able to get information from is very atypical of the Universe as a whole. We will then have a hard time finding the underlying Theory of Everything.

Barrow discusses the notion of the compressibility of information. A law of physics may be seen as a way to compress physical information in an abbreviated form. Belief in a Theory of Everything amounts to the belief that all physical processes can be rendered by one short formula, a formula 'you could wear on your T-shirt', as Paul Davies puts it. It is like representing all possible chess games by just stating the rules of chess. Barrow is certainly more modest here than is 't Hooft who would have the theory also fix exactly which particular chess game is being played in Nature. Among the various aspects of modern candidate theories that Barrow discusses I mention the notion of 'broken symmetry'. Symmetry amounts to saying that things that look different are actually the same. Physicists are fond of symmetries, for two reasons: first, because symmetry compresses information, and second, because it is beautiful. Beauty in physics is a very important but very dubious notion about which I will have more to say later. Unfortunately, the kind of symmetry that theorists would like their fundamental theories to have does not always manifest itself in physical fact. The symmetry is then said to be 'broken': it is present in the underlying theory, in the mind (of God?) so to speak, but not in our actual world. This compromise between what you want and what you have is typical of the way in which theoretical physicists manage to eat their cake and have it, and they sometimes get away with it surprisingly well.

After the reviews by 't Hooft and Barrow of the aims, problems and structure of modern fundamental physics, the philosophers take over and go right into the old debate between realists (McMullin) and empiricists (van Fraassen). It appears that the recent developments in physics are of no great importance to their outlook on physics. The realism–empiricism debate has a very long history and has become enormously complicated in our time. As McMullin puts it, there are almost as many realisms as realists, and the same can presumably be said of the anti-realisms. The general overview by Dennis Dieks is therefore most welcome. Roughly speaking, scientific realism is the belief that science deals with an independent external world and that the entities and structures that occur in our best physical theories 'really' exist in the external world. Scientific realism, in one form or other, is the position taken by 'all' working scientists. It is an essential part of their motivation. A scientist would be unlikely to spend a great deal of time and energy on discovering the properties of elementary particles unless convinced that they really existed 'out there'. There is a strange paradox here. For, although scientists regard realism as only natural, it is hard to define and defend philosophically. The most obvious argument in favour of realism is: if the entities that occur in our theories do not actually exist how is one to explain the extraordinary success of these theories? This argument has an inherent weakness, however, especially with regard to the theories that are claimed to be our most fundamental ones. For theories do not last forever. The impressive building of classical physics has been superseded by quantum mechanics, a conceptually totally different theory, and nobody will be surprised if quantum mechanics, in its turn, is eventually superseded by a completely different theory. Is it rational, then, to believe that in physics we are approaching the truth about the external world?

I seem to be entering the debate myself, whereas I should leave this to Ernan McMullin and Bas van Fraassen. The physicists will count on McMullin to defend realism for them. From his contribution it is clear that this was no easy task, and the realism that McMullin defends is remarkably restrictive. It is a 'local' realism: it applies to pieces of theory like the geological theory of plate tectonics or the chemical theory of molecular structure. What is real for McMullin is what can reliably be claimed to exist. Global physical theories like the gravitation theories of Newton and Einstein cannot easily be interpreted in realist terms because it is not clear what is being claimed to exist by these theories. This reluctance on the part of a defender of scientific realism to extend realism to these great physical theories will certainly not please those searching for a Theory of Everything. However, this is an important point: the more embracing physical theories are and the more they can explain, the less real their fundamental constituents seem to become.

McMullin and van Fraassen agree that a theory should be empirically adequate, i.e. it should be in agreement with the phenomena. But for van Fraassen this is also the only virtue that really matters: in the end, physicists will be willing to give up all other virtues they may have demanded of their theories, except this one. But according to McMullin physicists *do* require their theories to have other virtues, such as logical consistency, absence of *ad hoc* features, fertility or uniqueness, and he argues that this makes sense only in a realistic perspective. There are two deliberate omissions from his list of additional virtues: simplicity and explanatory power. These are left out because they are subjective and hard to define. If we equate simplicity with beauty, these are precisely the two virtues that theorists would rate most highly! Steven Weinberg, in his recent book *Dreams of a Final Theory*, devotes a whole chapter to the role of beauty in physics. To him beauty is part of what we mean by an explanation. We must conclude, then, that there is still a wide gulf even between a realist philosopher like McMullin and the theoretical physicists.

Even wider is the gulf between the empiricism of van Fraassen and the intuitions of theoretical physicists. According to van Fraassen acceptance of a theory does not require you to believe that it is true but only that it is empirically adequate. Empiricism may lead to an extreme scepticism about our possibilities of acquiring knowledge of the external world and, according to McMullin, this alone might be thought to constitute argument enough for some sort of realist alternative! So, how can one be an empiricist and still live happily? This is precisely the aspect taken up by Bas van Fraassen in his chapter, 'The world of empiricism'.

This beautiful article can be seen as the empiricist credo. It would be ridiculous to try and summarize it, but let me say a few words about it to connect it to the main themes of the symposium. Van Fraassen sees scientific realism as belief in a 'deep' structure of reality, to be revealed to us by our scientific inquiry. Knowing this deep structure would enable us to understand the world and to give meaning to it: it would give us a world-picture. For van Fraassen this realist view is pure metaphysics and he rejects it. He asks whether it would not be possible to do without this kind of world-view. He sees as the essence of science not the contents but the method. The value of science is not that it provides us with true theories, but rather that it accustoms us to changing our beliefs. He discusses the parallels between science and myth and ends his essay thus: 'What is the alternative to reifying the content of science? The alternative is to accept the challenge of intellectual maturity: let your faith not be a dogma but a search, not an answer but a question and a quest, and immerse yourself in a new world-picture without allowing yourself to be swallowed up. Science allows perfectly well the sceptical discipline that accepts appearances alone as real, and all the rest as a unifying myth to light our path.'

Contrast this view of a philosopher of science with the extreme realism of physicists like 't Hooft and Davies who virtually equate physics with reality. It seems to me that the views of the philosophical 'opponents', McMullin and van Fraassen, are very much closer to each other than to the realist views of most theoretical physicists. It is ironic, then, to hear many theoretical physicists strongly reject 'all' metaphysics. In fact, there is a marked return to realism among theorists after the long reign of positivist 'Copenhagen' philosophy in physics. This proves to me that a realistic attitude is somehow a psychological prerequisite for successful research in science. Most theoretical physicists will point to their successes as proof that they are not just making things up, and this appears to be a strong argument. One can, however, look at this differently. When theorists get stuck in their search for a unifying theory they muster all the creativity and imagination they have in order to find a way out. They are quite willing to sacrifice most of their former ideas as long as the new ideas are promising. After long or short periods of intensive effort they usually come up with a theory that they find more satisfactory. Now, one can assume that this shows that there is an ultimate 'best' theory which the physicists will eventually discover. Many theorists think this way! But one can also look at this process as showing the immense variety of mathematical structures that physicists are capable of inventing to suit their needs. The existence of this immense variety of structures makes it improbable that one of them is the right one. Which alternative one believes to be true is, I think, largely a matter of one's temperament.

The topic of world-views, which was brought up by van Fraassen, is central to Paul Feyerabend's chapters. His lectures at the symposium were delivered with characteristic fervour. Feyerabend is undoubtedly best known for his book *Against Method* (published in 1975) in which he discusses whether science is characterized by a fixed and well-defined method, the famous 'scientific' method, which singles out science from all other activities aimed at the acquisition of knowledge. Feyerabend passionately defends the view that no such scientific method exists, and that the only thing that can be truthfully said about scientific practice is that 'anything goes'. Although this infamous statement may have shocked philosophers of science, I don't think it has disturbed many scientists. But Feyerabend goes even further. He also maintains that science is not a totally rational affair, that science does and must contain irrational elements, and that in this respect it is no better or worse than other traditions like astrology and witchcraft. It will not come as a surprise, then, that Feyerabend argues that there is no such thing as a scientific world-view either. He does this quite convincingly by citing the views of a great number of scientists from different fields. I will not try and summarize his two chapters but I'd like to add a few remarks to three of the examples given by Feyerabend.

A scientist like Luria, who concentrates on 'small' well-defined problems which can be decided by clear-cut experiments, exemplifies the local realism of McMullin. Luria's attitude is, I believe, the one adopted by most experimental physicists.

A scientist like Einstein, on the other hand, exemplifies van Fraassen's realist who is searching for the 'deep' structure of Nature. The amazing success that Einstein achieved, almost single-handed, has had an enormous influence on theoretical physicists. Relativity theory was so all-embracing and convincing that it seemed almost impossible not to believe that it was here to stay and could never be changed. I think that the figure of Einstein and the success he achieved have, more than anything else, seduced theoretical physicists into believing that Nature has a fixed 'deep' structure, waiting to be discovered by a clever theorist, and that the way to discover it is to do as Einstein did: sit down at your desk and just think deeply. The views that experimentalists and theorists have of physics are really very different and I know experimental physicists who think that theoretical physics is not quite genuine physics!

Another effect of the wonderful success of relativity theory and of theoretical physics generally is that it makes one think that in the end the world *is* theoretical physics. Einstein once wondered whether God had had any choice when he created the Universe, or whether He was bound by the deep, rigid structures that theoretical physics had revealed. This comes close to identifying God with these structures, which is in fact what Hawking seems to do.

How different are the views of Wolfgang Pauli! Pauli was one of the most celebrated theoretical physicists of his time, a mathematical prodigy, feared for his tart criticism. To many people he was the essence of rationality. I was very surprised, therefore, when he published a book together with the Swiss psychiatrist C. G. Jung (*Naturerklärung und Psyche*, 1952) in which he applied ideas from Jung's psychology to the formation of physical theories by Kepler. Even among psychologists Jung is not regarded as a paragon of rationality and I could hardly believe that Pauli took him seriously. At that time I was very much interested in Jung, an interest I was careful to conceal from my physics colleagues so that they would not disqualify me as a future theoretical physicist. I gave a seminar on Pauli's article on Kepler and, much to my surprise, it was a great success. Until recently this article was all that the outside world got to know about Pauli's interest in Jungian psychology. The recent publication of his correspondence, showing his profound interest in the interpretation of his dreams and revealing his lifelong struggle to reconcile the rational and the irrational in his life as well as in science, came as a shock to many physicists.

Paul Feyerabend's conclusion seems very sound to me: science is consistent with many world-views.

In his first contribution, Bas van Fraassen discussed parallels between science and myth. In his second contribution, 'Interpretation of science; science as interpretation', he compares science with art. Both science and art represent Nature. But the concept of 'representation' is extremely complex and cannot be separated from interpretation. A genuine work of art can be interpreted in many ways. Every reader or spectator makes his or her own interpretation. A work of art does not spell out how it is to be understood. A scientific theory, on the other hand, seems to have only one interpretation, or, at least that is likely to have been its author's intention. Apparently, a theory that allows for more than one interpretation cannot be a good scientific theory. But according to van Fraassen this is a misconception: 'The history of science puts the lie to this story in successively more radical ways'. The chief example is, of course, quantum mechanics, which is both a remarkably good scientific theory *and* open to many interpretations. But even Newtonian mechanics has more than one interpretation.

Whether or not one allows a scientific theory to be open to more than one interpretation will, of course, depend on one's philosophical view of science (and of reality!) itself. To scientific realists the aim of science is truth, so to them a really good theory should not leave anything open; ambiguity is a defect. But to an empiricist a scientific theory need not be true to be good. As long as the empirical predictions of the theory are adequate the non-uniqueness of its interpretation is not a defect. On the contrary, each tenable interpretation casts new light on the theory and may point to a way of making progress.

Van Fraassen also discusses the notion of 'representation *as*'. Most physicists will agree, I think, that physics represents Nature *as* a mathematical structure. But physicists believing in a Theory of Everything may (and must?) go further. For them it is not enough for a final theory to represent Nature as a mathematical structure. The mathematical structure *is* Nature! 'If our universal law is sufficiently simple one could just as well conclude that the entire Universe is nothing but an enormous series of mathematical combinatorial "theorems"', says Gerard 't Hooft in his chapter. There seems to be a world of difference between this view and the view that sees science, like art, as an ever-open, never-ending exploration of the unknown.

When cosmologists tell us that the Universe came into being in the Big Bang, we may wonder how this ties in with the belief that God created the world. Do the results of modern cosmology have anything to do with

God, as Stephen Hawking suggests they might? The Big Bang theory represents our present-day knowledge about the Universe; what will happen to God if the theory turns out to be incorrect? It seems that only a final physical theory can qualify as a representation of the mind of God. Will *we* be special in the light of such a theory? And, is there any connection at all between a God who is revealed in the abstruse laws of physics and the God of good and evil to whom we turn for guidance and help? What kind of God are we talking about when we discuss 'God and the New Physics', and what kind of physics? The closed physics of the Theory of Everything, or the open physics of unimaginable developments? These and similar questions were the subject matter of the third part of the symposium, devoted to physics and theology. Willem Drees, who studied theoretical physics as well as theology and is the author of a book on God and modern cosmology, opened this last part of the symposium with an extensive general introduction to the subject.

Paul Davies has written a great many articles and several books on topics in theoretical physics but he is best known for his books that explain modern physics to a broad public. *The Mind of God* is one of his recent books. In his chapter, Davies explains the origin of the title. Western science emerged under the twin influences of Greek philosophy and the Judaeo-Christian religion. The Greek philosophers believed that we could come to know the Nature of the Universe through logical thought. The Jewish religion established the concept of a transcendent God who created the Universe and ordered it by imposing laws. These two strands of thought were united in the thirteenth century by Thomas Aquinas who conceived of a perfect and rational God who created the Universe as a manifestation of his supreme powers of logical reasoning. When Newton and his contemporaries formulated the basic principles of physics, they were convinced that their discoveries revealed God's handiwork, and that Nature's rational order had a divine origin. These scientists conceived of a Universe ordered according to definite laws of Nature, which they treated as 'thought in the mind of God'.

Davies gives a succinct account of what he sees as the most striking features of modern physics. The laws of physics have a very special form which facilitates the emergence of structures of great complexity from a featureless origin. No separate act of creation is called for. The Universe is self-creating as well as self-organizing. These features give the Universe the appearance of having a purpose. Davies calls this 'teleology without teleology'. According to him today's scientists are generally agreed that the fundamental laws of physics are universal, absolute, omnipotent and eternal, and many believe that they exist independently of the state of the physical Universe, and possibly independently of the Universe altogether. (I must confess that I was more than a little surprised by this statement and was forcefully reminded of what Paul Feyerabend has to say in his contribution on 'the' scientific world-view.) This deepest level of explanation could be called 'God'. Davies goes on to argue how strange it is that we humans are able to discover these eternal laws of Nature, or, as Heinz Pagels has put it, 'crack the cosmic code'. In Davies' opinion this ability points to a deep connection between our existence as conscious, thinking beings and the existence of the physical Universe with its bottom-level structure of basic laws and particles, and he sees the emergence of mind not as an incidental but as a fundamental feature of the Universe. He concludes his lecture thus: 'In some strange and perhaps unfathomable manner, it begins to look as if we are meant to be here.'

The final chapter, 'The sources of models for God: metaphysics or metaphor?', by Mary Hesse, can be read as a direct reply to Paul Davies. Hesse does not share Davies' enthusiasm for a Theory of Everything. She calls herself a 'moderate' realist: we are always in local and particular situations, however far into the past or future or into micro- or macro-scale our theories claim to penetrate. Science has been successful in ordering the, always local, natural environment to permit prediction and control. But the local and particular success of science does not require there to be true *general* laws and theories.

One could perhaps use here the metaphor of the scientist as someone who holds a candle to illuminate the surroundings but does not know how far the darkness extends. (By the same analogy a believer in a Theory of Everything would believe that there is a light switch somewhere and that we are close to finding it.)

But even if the mathematical structure that we are uncovering is the Theory of Everything, it is not strong enough to serve as a model for

God. For it lacks precisely those qualities which touch the religious issues of everyday life. It introduces no values such as goodness or freedom. Hesse goes on to discuss other models of God besides cosmological ones. She indicates the sort of arguments that may be used for and against them, and finds these arguments quite different from those deployed in science. Her conclusion is that these models should all be understood as metaphors for God, not as metaphysical descriptions of God.

In a very interesting final section, Hesse compares models of God with models in physics. To a moderate realist physical models, like models of God, are models of something the essence of which we cannot capture. The only thing we know about natural reality is that it constrains the outcomes of our experiments. But this does not imply that we are forbidden to speak of, or refer to, a natural reality. The ontological belief that there is a constraining natural reality does not depend on one's ability to represent it accurately. We must speak about God *and* about natural reality in metaphors, but this does not preclude belief in their existence.

This compactly written article, which also touches on the problem of theological language, requires close reading. Fortunately, Mary Hesse's careful replies to questions in the discussion section provide valuable further elucidation.

Summarizing this summary, and generalizing a little, one finds the three physicists united in their enthusiasm for a Theory of Everything, although they differ somewhat on what it means. They also make it quite clear that we may never be able to find such a theory or make practical use of it. The philosophers among the contributors are less inclined to attach great importance to the idea of a final theory. Whereas physicists tend to see the world as a manifestation of the laws of physics, the philosophers see physics as one way among many others to acquire knowledge about the world. The physicists are realists about the laws of physics; to them these laws represent or even 'are' the world as it really is. The philosophers look very differently at this. Their views range from a moderate, 'local' realism to an outright anti-realism (appearances alone are real). The physicists stress the amazing potential of the current physical laws to produce structures of great complexity starting from very simple beginnings and they point to the (hoped for) uniqueness of the deepest laws. The philosophers see current physics as just a phase in a potentially endless process of expanding scientific knowledge. In a Theory of Everything there is no place for God, except as a name for this fundamental mathematical structure. Such a model for God has little to do with what most believers would call God.

At this symposium we were privileged to hear the views of a number of physicists and philosophers who are foremost in their fields. It will be clear from the contributions contained in this book that there is no consensus, no uniformity. Our knowledge, however extensive it may be, remains open to many interpretations.

## Acknowledgements

I wish to thank Dennis Dieks, Fred Muller and Jos Uffink of the Department of History and Foundations of Mathematics and Science of Utrecht University, and Hanneke Pasveer, secretary, for their help in preparing the symposium. Special thanks are due to Bas van Fraassen for suggesting speakers and to Willem Drees for his unfailing help on the physics and theology section. I should also like to thank Joop Doorman, Gerard Nienhuis and Vincent Brümmer for acting as chairmen at the symposium. Yvonne Goester and Adra de Jong of the Praemium Erasmianum Foundation performed the demanding task of producing a typed version of the recorded discussions from which Christel Lutz and Heinerich Kaegi extracted a list of questions and replies. I am indebted to Meta Hilgevoord and Jos Uffink for their critical remarks on an earlier version of the introduction to this book. I am extremely grateful to Sheila McNab; at all stages I could depend on her for conscientious help with the English. Finally, I greatly enjoyed cooperating with Hans Hoetink, director of the Praemium Erasmianum Foundation and his staff members Yvonne Goester and Adra de Jong.

# 1 Questioning the answers or Stumbling upon good and bad Theories of Everything

Gerard 't Hooft

**Summary**

The last two decades witnessed a new breakthrough of fundamental physics in its struggle towards a better understanding of the World of the Small. This gave rise to new speculations concerning the idea that there may exist a single ultimate Law of Physics underlying all particles and forces, and therefore also everything made out of these particles held together by these forces, which could include us and the universe. Discovering this Law would be a tremendous achievement, but would not at all imply our understanding of most of its consequences. In this chapter I focus on two themes:

(1) Speculations that such a fundamental Law exists and that there is a possibility that we humble human beings may in due time be able to find this Law and give a detailed and precise formulation of it – including boundary conditions and initial conditions if necessary – are far from ridiculous (Section two), in spite of lessons from the past. Our present knowledge indicates that the possibility is real (Section three).

(2) Claims made by several theoretical physicists that they are coming very close towards actually realizing this hope are highly premature and naive. 'Superstring theory' is almost certainly not the answer. The dispute concerning the physical interpretation to be given to the formalism called 'quantum mechanics' will have to be continued, and it is as yet impossible to predict the vastness of the undoubtedly formidable obstacles that are still in our way (Sections four and five).

Practical applications of our allegedly expected understanding of a Universal Law will be very limited and will certainly have no effect on our daily life.

But philosophical implications will be far-reaching. I will try to explain what I mean by the complete merging of theoretical physics with mathematics.

### 1 What could possibly be the meaning of the words 'Theory of Everything'?

Science, as we know it at present, is based on empirical knowledge. In our perpetual quest for a better understanding of the patterns and forces that govern our world, we have learned how to devise all sorts of machinery for making subtle observations, and to perform experiments by which theories concerning these patterns and forces can be checked. Experimental observations are never perfect, and theories can never be absolutely confirmed. In the best of circumstances a theory can either be falsified or given further support. In view of this it would seem to be silly to ever expect that we can brew a *perfect* theory. There is an infinity of phenomena still waiting to be studied by us, and so one should expect also an infinity of theories, some better than others, but none perfect, let alone *one single* theory that could accommodate everything at once. If the words 'Theory of Everything' were to be interpreted literally it would be quite obvious that such a concept will never be fabricated by mortal human beings.

What one can speculate about in a more fruitful manner is something that could be called an 'Ultimate Law for Basic Dynamics'. What I mean by this is best illustrated by formulating an example for a basic Law of Dynamics that could be used to generate an entire universe. The example I have in mind is an idea that originated in the early 1970s when physicists and mathematicians began to play various apparently nonsensical games with computers. When the first personal computers arrived somewhat later, it became very popular. The idea was called 'Conway's Game of Life', and it went as follows (see Fig. 1).

On a rectangular infinite lattice we have 'cells', each of which carries one 'bit' of information: the cell is said to be 'alive' if this bit of information is a one; it is 'dead' if it is a zero. Then there is a clock. At every tick of the clock the contents of each cell is being updated. For each cell the new status, at time $t + 1$, depends on the contents of itself and its nearest eight neighbors at time $t$ (Fig. 2).

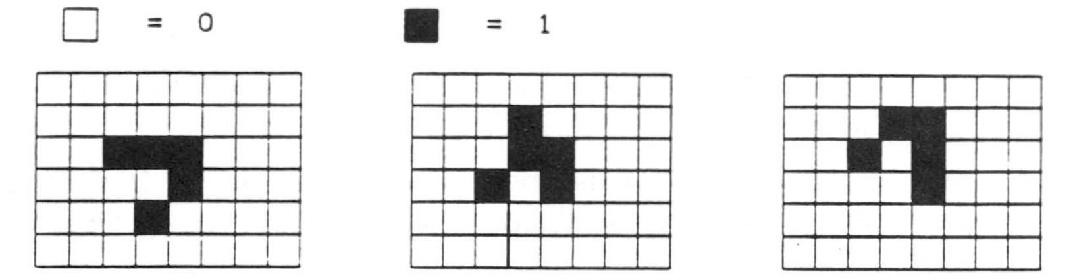


Fig. 1. Evolution of a particular pattern in Conway's Game of Life, at three consecutive times. It will propagate diagonally over the lattice.

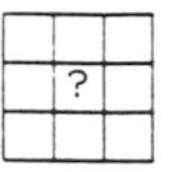


Fig. 2. The cell in the center is updated depending on what was there before, and on what was in the eight surrounding cells drawn here.

The rule is as follows:

- *If exactly two neighbors are alive, the cell in the center will stay as it was.*
- *If exactly three neighbors are alive, the cell in the center will live.*
- *In all other cases the cell in the center will die.*

In Fig. 1, I show a pattern that, after a while, returns into itself, but at a different place: it moves! One can start with an initial configuration where several of these moving patterns are sent towards each other so that they will collide, and then study what happens. The results may become rather complex, and without a computer hopeless to calculate.

The point I want to make is that this system is a 'model universe'. Imagine that, given a large enough lattice and a patient enough computer, an 'experiment' is run for a sufficiently long time. Regular, recognizable structures may appear again and again, satisfying their own 'laws of physics'. In principle these laws of physics should be derivable from the original rules mentioned above. We could call these structures 'atoms'. Atoms themselves will form 'molecules', and so on. Eventually, one might find 'intelligent' creatures built out of these building blocks. We might call them 'humans'. They will investigate the world they are in, and perhaps ultimately discover the three fundamental 'Laws of physics' on which their universe is based.

It is unlikely that the rules given above will do all that. But the question one may nevertheless ask now is: could it possibly be that the universe we are in ourselves has a similar simple structure, based on such a simple universal Law, a Law that is invariable and absolute? Is it conceivable that we will ever be able to discover that Law if it exists? The first thing one notices is that, although Conway's world is at best only a poor caricature of our real world, there are striking similarities with the real world. Conway's world does contain different kinds of 'atoms', of which the moving pattern shown above is only one example. There are many more. These particles can collide against each other and interact. Furthermore, something else is needed before a complete description of the Universe is obtained: one should also formulate the *boundary conditions*. Not only should we state how large the lattice is, and whether it closes into itself like a sphere or a torus (a doughnut shape); one must also specify what the *starting configuration* was: the boundary condition at time $t = 0$. There are even versions of models of this sort that can describe an 'expanding universe', much like our own. The boundary conditions may then be extremely simple and yet the details of the structures obtained at later times may be extremely complicated.

It is interesting to speculate that in a model of Conway's type various particles might form super structures such as 'atoms' and 'molecules', and all large scale structures one thus obtains, and all their properties, will be sole consequences of the single three Laws that I formulated in the beginning. Just imagine that a community of 'intelligent' creatures could emerge in a Conway-like system. By patiently performing numerous series of experiments, constructing and rejecting one theory after the other, these creatures might eventually stumble upon the conjecture that their universe is generated by just three basic Laws. All of their experiments will be in agreement with these Laws, although these creatures can never be sure that the formulae they found are the ultimate ones. It could always be conjectured that exactly once every 1 048 576 ticks of the clock the rule is different: if the closest 624 neighbors are all in one particular given pattern you get a one instead of a zero. But after elaborate tests one could rule out such proposals one by one, and a moment will arrive when most of the inhabitants will be convinced that the Ultimate Law has been found.

Of course, by guessing the Ultimate Law not all phenomena in Conway's world would be immediately explained. In fact it is only then that at last physicists could begin explaining the phenomena they see one by one, and the ultimate 'explanation' of phenomena as complex as life, let alone intelligent life, will almost certainly never come. It is an Ultimate Law that physicists are seriously thinking about, not a theory of everything in the literal sense. The name 'Theory of Everything' instead of 'Ultimate Law' or 'Basic Law' was a bad commercial.

In summary, what I would like to discuss here, and what I will indicate admittedly inaccurately by the misnomer 'Theory of Everything', is a set of laws of physical dynamics with the following properties:

(1) The laws will determine with infinite accuracy the evolution of all physical dynamical variables at a local level, and should also include a description of the 'boundary' of the universe, as well as its initial state.
(2) There exists no closely resembling alternative theory. This means that any slight change brought about in the rules would make the theory unlikely or inelegant. The theory will be a 'package deal': take it, or leave it. This should hold both for the local laws and for the boundary conditions.
(3) Evolution according to these laws will give rise to a nearly infinite complexity, a complexity sufficiently extensive to include the marvelously perplexing wonders abounding in our universe – the emergence of life and intelligence being only a few of these.

The phenomenon of Complexity is familiar among physicists. A well-known example is the weather system in our atmosphere. In combination with another complex phenomenon, namely the formation of continents on Earth, it gave our planet the enormously rich structure it has. But for non-physicists it is often difficult to imagine how even much simpler basic laws can sometimes produce equally or even more complex behavior. I believe that the reason why many of its opponents intuitively reject the notion of a theory of everything is that they underestimate the effects of complex behavior. More about this in Section five.

## 2 The present theory of elementary particles and forces

How does our present understanding of elementary particles compare with a 'Theory of Everything'? In certain respects we have made great progress in precisely that direction, but on the other hand it is quite clear that the gaps separating us from any 'ideal' theory are still tremendous. Let me discuss these two observations in turn.

In the 1970s a development took place in quantum field theory that allowed us to formulate a model for all known particles and forces between them. This model became known as 'The Standard Model'. It is a set of equations which were found to give a remarkably accurate description of precisely all the particles and forces ever detected in any laboratory experiment. The model is fairly complicated, and it cannot predict the reaction of particles and fields under all conceivable circumstances – after all, in our fantasy we could easily imagine experimental setups that are completely impossible to realize in practice. But it seems to be a very good description of what is happening nearly everywhere in our universe. I will now give a rough description of this model and its laws.

The basic entities are *Dirac fermions*, elementary particles that show a certain amount of spinning motion. This amount of spinning motion is measured in multiples of a fundamental constant of Nature, Planck's constant divided by $2\pi$: these particles are said to have 'spin $\frac{1}{2}$'. We start with particles that can only move with the speed of light. For these particles the *axis of rotation* can be seen to be always parallel to their velocity vector, and one can derive that this statement is unique and independent of the velocity of the observer. This is why one can distinguish unambiguously particles that spin 'to the left' from particles that spin 'to the right', with respect to this axis. From a mathematical point of view there is some resemblance between particles of this type and the zeros and ones in Conway's model.

The forces these fermions exert on each other are now described by introducing another set of particles, the *gauge bosons*. These are particles with spin 1. They can either be considered as the 'energy quanta' of various kinds of electric and magnetic fields, or one can view these particles themselves as the transmitters of these forces. When that picture is used one describes the force as being the consequence of an *exchange* of a gauge boson between two fermions: one fermion emits a boson and the other fermion absorbs it. If the mass of the boson is negligible then the efficiency of this exchange process is inversely proportional to the square of the distance: the Coulomb force law. If the boson has a certain amount of mass when at rest, then the force it transmits will range only up to a distance inversely proportional to that mass, and decrease very rapidly beyond that distance.

An important feature of the gauge boson force between two fermions is that before and after the emission of a gauge boson the fermion keeps the *same helicity*. This means that a left rotating particle remains left rotating, and a right rotating particle remains right rotating. This is of special importance for the neutrinos: neutrinos *only* exist as left rotating objects. Right handed neutrinos have never been observed (at least not for sure). In contrast, *anti*-neutrinos only come rotating towards the right, not left. By emitting a gauge boson a neutrino may turn into an electron, but then the electron must rotate towards the left.

We can now make a listing of all fermions and gauge particles (see Fig. 3). Here we see that the fermions are divided into *leptons* and *quarks*. The left rotating objects are indicated by an '*L*' and the right rotating ones by an '*R*'. For the anti-leptons and anti-quarks,

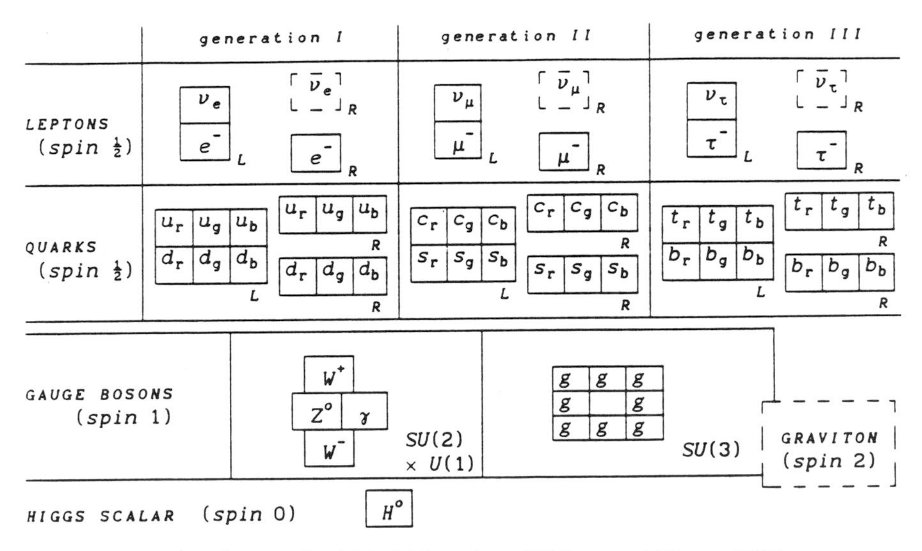


Fig. 3. The Standard Model based on $SU(2)_{\text{weak}} \times U(1)_{\text{em}} \times SU(3)_{\text{strong}}$. Right handed neutrinos, which may exist, are indicated in dotted boxes.

which are not shown, the *L* and *R* are interchanged. Of all fermions only the quarks are sensitive to the forces of the eight gauge bosons in the box called '*SU*(3)'. This force is very strong and so all objects containing quarks will be strongly interacting with other objects.

All *left* rotating fermions are sensitive to the forces from the *SU*(2) gauge bosons. Finally, all left rotating, and all electrically charged right rotating fermions feel the *U*(1) force fields.

Now the Standard Model also contains a spin 0 particle, the *Higgs particle*. It also transmits a force, but an important difference between the latter and the gauge boson forces is that if a fermion exchanges a Higgs boson it has to flip from left rotating to right rotating or vice versa. Another difference is that the spin 0 particle can disappear straight into the vacuum. This implies that some fermions can make spontaneous transitions from left to right and vice versa. Technically, this is the way one may introduce *mass* for these fermions. A particle with mass moves slower than the speed of light and it turns out that for such a particle the rotation axis cannot be kept parallel to the velocity vector. So this mass can only be there if the particle has the ability to change its spin direction relative to its velocity vector. One finds that the mass of a fermion is proportional to the coupling strength of that fermion to the field of the Higgs particle.

I should emphasize that this theoretical understanding of the behavior of fundamental particles in terms of relatively simple equations is due to a huge combination of efforts by theoretical and experimental physicists alike. Without numerous ingenious experiments, sometimes carried out in very sober conditions but often in enormous multinational collaborations in gigantic laboratories, the present insights would perhaps never have been obtained.

The *gravitational force* could be added to our description of the Standard Model by postulating a spin 2 particle, the *graviton*. The rules for this force are very strictly prescribed by Einstein's General Theory of Relativity, but only in as far as it acts collectively on many particles. The details of multiple graviton exchange between individual particles are not understood, mainly because any effects due to such exchanges would be so tremendously weak that no experimental verification will be possible within any foreseeable future.

The above is a qualitative description of the rules according to which the particles in the Standard Model move. To formulate these rules in more precise mathematical terms, as we were able to do for Conway's Game of Life, would lead us way beyond the scope of this chapter. In some sense the rules are nearly as precise.

Allow me to point out the similarities between this Standard Model and an ideal Theory of Everything: all material objects in the Universe have to move according to its rules. In some sense the standard model is nearly as esthetic and elegant as Conway's model. But of course the standard model is far from perfect. One reason is that we expect the model not to provide the equations of motion for matter under all conceivable circumstances. One can imagine energies, temperatures and matter densities that are so large that the situation cannot be mimicked in any accelerator. Although our model could in principle be used to calculate what will happen, there are several indications that one should not take such predictions seriously. A second reason is that the *mathematics* of the model is less than ideal. Even if we had infinitely powerful computers at our disposal, some phenomena cannot be computed with any reasonable precision. One has to rely on certain perturbative approximation schemes that sometimes fail catastrophically.

One of the most significant distinctions between the present Standard Model and a Theory of Everything as defined in the previous section is that our interactions depend on a number of 'constants of Nature', which are to be given as real numbers. The precision with which these real numbers are known will always be limited. There are essentially 20 independent numbers:

- 3 gauge coupling constants, corresponding to the strength with which the various kinds of gauge bosons couple to the fermions. The $SU(3)$ coupling constant is much larger than the one for $SU(2)$ and $U(1)$. Then there are several interaction terms between the Higgs field and the fermionic fields (Yukawa terms). Many of them correspond to the masses of the various fermions:
- 3 lepton masses: $m_e$, $m_\mu$ and $m_\tau$.
- 6 quark masses: $m_u$, $m_d$, $m_c$, $m_s$, $m_t$ and $m_b$.
- 4 quark mixing angles, determining further details of the decay of exotic particles.

- 1 topological angle $\theta_s$, a peculiarity relevant only for the strong interactions; as far as is known it is very close to zero.
- 2 self-interaction parameters for the Higgs field. One of these determines the Higgs mass $M_H$, the other determines the Higgs-to-vacuum transitions. In combinations with the other constants it produces the gauge boson masses.

This adds up to 19 'constants of Nature', which are incalculable; they have to be determined by experiment. We could then add Newton's gravitational constant $G$, but this could be used to fix the as yet arbitrary scale for mass, length and time. Strictly speaking there is also the so-called *cosmological coupling constant* which is also incalculable, but it may perhaps be set to be identically zero; it would be the 20th parameter.

The presence of these constants of Nature violates condition (2) for a Theory of Everything to be acceptable. This is because with this model we will always have the option to consider an arbitrarily tiny change in one or more of these numbers, and thus come forward with an 'alternative theory' that is equally as probable as the original one and cannot be distinguished from it by any experimental test. Experimental tests always have limited precision. A good Theory of Everything should not contain any freely adjustable constants of Nature that take the form of real numbers.

There must be three constants of Nature that are not adjustable and therefore acceptable in any theory. These are the *scale determining constants*, for which we usually take:

– the speed of light, $c = 2.997\ 924 \times 10^8$ m s$^{-1}$
– Planck's constant, $\hbar = h/2\pi = 1.054\ 588 \times 10^{-34}$ J s
– Newton's gravitational constant, $G = 6.672 \times 10^{-11}$ m$^3$ kg$^{-1}$ s$^{-2}$

Together, these determine a fundamental unit of length,

$$\sqrt{\frac{G\hbar}{c^3}} = 1.6 \times 10^{-35} \text{ m}$$

one unit for time,

$$\sqrt{\frac{G\hbar}{c^5}} = 5.4 \times 10^{-44} \text{ s}$$

and a unit for mass:

$$\sqrt{\frac{\hbar c}{G}} = 22\,\mu\text{g}$$

with a corresponding energy of $1.2 \times 10^{22}$ MeV (or $2.0 \times 10^{9}$ J).

These units are called the Planck length, Planck time and Planck mass or energy. One could use these to dispose of the arbitrariness of the 'meter', the 'second' and the 'kilogram' as units for length, time and mass. In terms of the Planck units the three corresponding constants of Nature are just one, and therefore not adjustable.

Just because the gravitational constant is extremely tiny at the level of elementary particles, these resulting 'natural' time and length scales are tremendously small, and, compared to particles, the mass is extremely large.

It is for structures at these distance, time and mass scales that our Standard Model is known to be desperately inadequate. We know that our present formalism of quantum field theory is not suitable to describe the gravitational force among individual particles. What would be needed is a theory that combines our present knowledge of both quantum mechanics and Einstein's general relativity into one scheme, and in spite of vigorous attempts from theorists such a scheme has not been found. It is generally believed that the present formalisms all have to be completely revised before they can be combined. We do not know how this can be done, and this is why there will be room for speculations.

The Standard Model is elegant and esthetic, but not elegant and esthetic enough; it is universally valid, but not universally enough, and so on. And then, still very little is known about possible boundary conditions of our universe. The science of cosmology, that should tell us the answers to these questions is making progress, but should still be regarded as being in its infancy.

## 3 Is there any evidence for a universal law?

The evidence in favor of the existence of a universal law was clearly visible to the physicists of the previous century. If Conway's model and personal computers had existed then, they would immediately have pointed to the striking similarities between that model and the real world. Only religious arguments or mystical reflections could be raised against the idea, and it is to be suspected that most scientists would reject such counter arguments. Anyone with some experience in the universal nature and mathematical strictness of the laws of physics would be very much tempted to believe that there exists a completely general scheme according to which everything moves.

And there are now also elements in the most recent theories that lend further support to this view. The existence of the Planck length and time scale suggests that at that scale the laws are fundamental and absolute. The requirement that these laws must agree with the very restrictive postulates of both quantum mechanics and general relativity has up to now proved to be so difficult to realize in any physical model that one is tempted to suspect that not more than one model will exist at all which agrees with all this. More significantly, studies of the quantum mechanical laws in the vicinity of black holes indicate that the total amount of *information* one can store in any given volume is limited. This is precisely what we also have in Conway's model!

The idea of a Theory of Everything was given further new impetus in 1984 when a new theory of particles was formulated that includes the gravitational force. It became known as 'superstring theory'. This idea became very popular because it seemed to obey requirement (2): it contained no freely adjustable constant of Nature. Unfortunately, the mathematics could never be given such a concise form as in Conway's model, and at present a majority of particle physicists have lost most of their high expectations of this scheme. String theory is a mathematical device, not a complete theory of particles and forces.

We can play with toy models of an ultimately mathematical universe, but we cannot ignore an overwhelmingly important piece of evidence *against* the existence of any local law of dynamics. It is an aspect of the theory of quantum mechanics itself. When the fundamental laws of quantum mechanics became known, in the 1920s, it was also realized that there is an enormous disparity between this theory and any system based upon local and deterministic dynamical laws. By 'local' we mean to say that the evolution of some physical quantity should only depend on physical quantities in its immediate neighborhood, so that there is no action at a distance. Action at a distance was already excluded by Einstein's theory of relativity: interactions over some distance can only occur if some sort of 'field' propagates through the region in between. 'Deterministic' means that one single initial configuration can lead to one and only one possible configuration later. If locality and determinism were fundamental requirements of a Theory of Everything then quantum mechanics would be at odds with such a theory. And the evidence that quantum mechanics correctly describes all tiny objects such as atoms, molecules and all other small particles, is overwhelming.

In contrast, according to the laws of quantum mechanics any initial configuration may lead to many different possible outcomes, each with a precisely calculable probability. An atomic physicist equipped with the theory of quantum mechanics can never predict all details of the outcome of experiments, but only probabilities for many different outcomes, as if throwing dice.

Attempts to remedy this situation, replacing quantum mechanics by specially constructed deterministic models containing some statistical elements, were unsuccessful. In fact, it could be *proven* that many such models fail to reproduce the statistical predictions from a quantum mechanical calculation. Einstein, Rosen and Podolsky constructed a 'Gedanken' experiment of which quantum theory gives a precise prediction (after an easy calculation) that cannot be reproduced unless one has action at a distance. Bell later formulated rigorous mathematical theorems with the same outcome.

In my opinion (but I stress that this is a minority view), there may nevertheless be a compromise. This is that there is no direct action at a distance, but there is some sort of 'conspiracy'. With this I mean that the 'state' of Nature that we now call 'vacuum' is actually a very complicated dynamical solution of the equations of motion, showing correlations over space-like distances. Einstein, Rosen, Podolsky and Bell never took such correlations completely into account. With correlations one can have apparently impossible 'coincidences' spreading faster than the speed of light, but which are not in conflict with the requirement of special relativity that *information* cannot spread faster than the speed of light. You may ignore this deviating view because it is not essential for the main theme of this chapter. I will come back to it, briefly, later.

More in line with standard interpretations of quantum mechanics would be the idea that we might still have something resembling Conway's Game of Life, at the Planck scale of time and distance, but that it is a 'quantum mechanical' version of the model. A slight complication might be that the number of 'quantum-Conway' models seems to be much larger than that of classical, deterministic ones. On the other hand, requiring such a model to include the postulates of general relativity may, as stated earlier, reduce the number of possible schemes to perhaps only one. Today, no such model is known at all.

## 4 What are the prospects of discovering the basic laws?

No working model that combines all we know at present concerning the basic laws of physics is known at all. Yet there is ample evidence that Nature itself did produce such a model, namely our present universe. So we should not despair: a mathematical scheme exists; it just remains to be uncovered by us. Will we ever succeed in doing this?

The lessons from recent and more distant history are instructive. A number of times in the past physicists have expressed as their opinion that the ultimate laws of the world had 'nearly' been found. They were always proven wrong. Is it not evidently more reasonable to expect only piecewise improvements in our understanding of Nature's secrets, and that the series of false alleys and surprises awaiting us will be strictly endless? Or that progress will eventually slow down and come to an end just because we will never be able to afford the formidable expenses needed to build the next generation machines, before enough information were obtained from which we could deduce the basic laws?

Yes, it may be feared that history will repeat itself, but history never repeats itself in a predictable manner. So the fact that claims in the past turned out to be premature does not imply that future claims of this sort will be premature also. The recent claim by superstring theorists that the Theory of Everything was close at hand enjoyed the echo of a considerable number of cheer-leaders, but I did not belong to them. This theory did not have the characteristics of any basic theory that would be acceptable to me. The mathematical formalism was far too complex and indirect to serve as a universal law. It was and is not understood how to work with this theory under all circumstances, and such is an absolutely obligatory requirement for any theory that boasts to be an ultimate one.

So, no, I must emphasize, we are not even close to the ultimate theory at present. We should keep in mind that the obstacles waiting for us are indeed formidable. The Planck scale corresponds to an environment where particles feel each other's gravitational forces. These forces under normal conditions are extremely feeble. Experiments in which these forces dominate the outcome of the measurements will probably never be possible. It is not unlikely that improvement over the present concepts can gain us a few orders of magnitude, but to raise the energies and/or accuracies of the measurements by 16 orders of magnitude is, it seems, a hopeless task. Information concerning this area of physics will have to be obtained by extrapolation, making optimal use of our theoretical and calculational ingenuities.

It is our human ingenuity on which I am counting. Ultimately I believe that clever ideas will be found enabling us to deduce how Nature works. If a simple fundamental law exists, it will be found, sooner or later. If the law is found, we will all recognize it as being very likely the truth. I have no doubt there. I also have no doubt that regardless how convincing the evidence for a new super theory will be, people will continue to raise questions and objections. There never will be an ultimate proof of the ultimate theory.

## 5 Aspects of complexity

It is very clear that objections can be raised against the very notion of a single all-embracing theory for the tiniest dynamical units in this world. The idea of one single and simple mechanism that is universally valid might be a complete illusion never to be applicable to the real world. I am not blind to the valid objections. Many people however may raise intuitive objections which one can put aside. One objection that I can imagine would be that a dynamical law as simple as that of

Conway's model could never give rise to all the riches of the real Universe, its immense beauty, diversity and magnitude, in particular the emergence of life, let alone intelligent life.

This is clearly an objection against the idea that simple initial laws could give rise to infinite complexity. In the lecture I paid little attention to the notion of complexity, for lack of time (I refer to John Barrow's chapter), but let me dwell on it, a little, in this written version.

Complexity is a quite common feature both in physical and in mathematical systems. A simple physical example is 'chaotic behavior', displayed in many systems that themselves are described by simple and straightforward laws. Consider a gas such as air. The laws for the motion of tiny volumes of air, the so-called Navier–Stokes equations, are fairly simple. To understand the behavior of air one may often ignore even the fact that it consists of molecules. We may treat it as an 'ideal' gas. Yet air may give rise to *turbulence*. Given a simple shape such as a perfect sphere, one can calculate, and also observe directly, how air flows around it. Under certain conditions the solutions of the relatively simple equations become extremely complicated. That is when turbulence shows up. We call this phenomenon 'chaos'. The complexity may very quickly reach the very limits of even the most powerful computers, and so, even if we have the full equations, there will be uncertainties when we try to apply them. The reason is not only that we do not know the equations precisely enough, but also that we can never control the initial state precisely enough, and that our mathematical formulations would never be sufficiently accurate, and finally that there are many more physical degrees of freedom than we can handle, even with the most powerful computers. Nature itself will always act as a computer much more powerful than anything we will be able to construct to mimic her.

This fact will imply that the questions we are addressing here, and whatever answers we come up with, will have little effect on everyday life. A Theory of Everything of the type we are discussing here will never be a theory of everything in the literal sense. Chaotic behavior will prevent us from computing just 'anything we want to know' at a very early stage. The question we are addressing is the question whether or not there exists an ultimate law and whether we will be able to discover this law. The question is not whether we will ever be able to use this law to explain whatever we like to have explained. Almost certainly, our dynamical laws will be useless for that, due to the tremendous amount of complexity, presumably already at scales a little bit larger than the Planck scale.

Conversely, one may easily use these observations to counter the objection that our universe is too complex to be described by one single dynamical law. We know that single dynamical laws may show infinite complexity in their solutions. Anyone who can program a computer can demonstrate this for him- or herself. Take a cellular automaton such as Conway's Game of Life, or some variation on this scheme. Under favorable conditions one can start with an extremely simple law and an extremely simple initial state, and find that its behavior grows infinitely complex.

Not only physical models, but even purely mathematical systems can easily become infinitely complex. One example is the series of prime numbers, which at extremely high orders becomes more and more difficult to predict. Another one is the famous mathematical function called the 'zeta function', of which some mathematical aspects are not completely understood because of its erratic behavior in the complex plane.

The beauty of our real world is that there is some equilibrium in its degrees of order and degrees of complexity. In my personal theory for the interpretation of the laws of quantum mechanics, it is complexity that dominates the structure of the 'vacuum', yet this complexity is far from random. There are correlations, and the only correct way to describe these is by what is now known as quantum mechanics. The laws of quantum mechanics as we know them contain a statistical element, just because the complexity of the vacuum is far too immense to allow it to be described in any other way.

I have not been able to explain why these vacuum correlations are stable and why the vacuum has the symmetry properties that we observe in practice, nor have I been able to reproduce all these properties in mathematical models, but I don't see why the general idea should be wrong. It could be that there exists only one such model, the one that describes the real world . . .

## 6 Just suppose . . .

Allow me for a short while to speculate that there is indeed a simple dynamical law, waiting to be discovered. As I explained and emphasized earlier, the practical implications would hardly be noticeable. It is almost certain that such a law will rapidly produce chaotic behavior, so that even the most elementary calculations using this law as a starting point will be cumbersome at best, impossible in most cases.

But from a more philosophical viewpoint the implications would be immense. We would have to conclude that what we used to call 'Nature' is actually something like a mathematical processor, much like the processors we have inside our computers, just quite a bit bigger and faster.

It is not obvious that Nature's own mathematical processor works just like ordinary ones. Most physicists would argue that, since we have quantum mechanics, Nature's processor will have to obey 'quantum logic' instead of ordinary logic. This is a widespread belief which, as I stated earlier, I do not share. What physicists at present call 'quantum logic' may well be nothing but the best representation of our present understanding of the laws of physics, but may not necessarily be the ultimate truth. From a philosophical point of view it could not have been otherwise than that our first attempts at formulating the laws of physics should contain statistical elements; all those dynamical degrees of freedom or variables that we do not understand at present just look like random noise to us. Experts will no doubt recognize in these words the 'conjecture of the hidden variables'. Since several simplistic versions of such theories have been ruled out in the past, the notion of hidden variables is no longer very popular in theoretical physics. So I will try not to emphasize my personal belief in some more advanced version of hidden variables, but I won't hide it either.

Whatever the logic, quantum or classical, random or deterministic, let us suppose that there exists a well-formulated physical law – and even that is not an obvious fact at all. Then I claim that there should also exist some prescription concerning the order in which the law –

or laws – should be applied to calculate how the numerous physical degrees of freedom evolve. It is this order that I would identify with 'time'. One must *first* calculate how 'early' degrees of freedom evolve, and only *then* one has the necessary data to calculate how 'later' degrees of freedom will behave. Without a good prescription of the logical order the laws will not be unambiguous. This inevitable fact seems to me to imply the necessary existence of *time* and *time ordering*. Going backwards in time, or following 'closed timelike loops' will be contradictions in terms.

I can even go a step further. If our universal law is sufficiently simple, one could just as well conclude that the entire universe is nothing but an enormous series of mathematical combinatorial 'theorems'. The mathematical theorems are time ordered. Time ordering is indeed also the order of logic: first come the theorems that were relatively easy to prove, then come the theorems that require the previous theorems to be proven first. Now many mathematical theorems are in some sense 'equivalent': they require tremendously complex calculations for their proofs, but one theorem follows directly if we know the other. For the universe this could mean that there may exist many equivalent descriptions of its dynamics.

A simple example of an infinite series of 'theorems' is the list of prime numbers. To prove that a large number is a prime number you first need to know the smaller prime numbers. Thus, the series of prime numbers in some sense form their own universe – indeed, one can represent the prime numbers as a four-dimensional table just like our real universe! If mathematicians could prove that all combinatorial theorems in number theory ultimately are connected to knowing the complete series of prime numbers, then our universe 'is' the series of prime numbers! This is what I referred to in the summary as the complete merging of theoretical physics with pure mathematics.

A very important problem then remains: the problem of identifying our present position in this universe! Present theories of cosmology indicate that galaxies should have formed from initial tiny fluctuations in a very early, very hot universe. These fluctuations will be extremely difficult, if not impossible, to calculate. One can imagine in the distant future a computer program that works out the equations and finds the rough shapes of the galaxies. Maybe we will recognize our own!

I mention this admittedly rather crazy idea as an introduction to something more important: the so-called 'anthropic principle'. According to this principle the Universe 'is as it is' because 'if it weren't, we would not be here to discuss it!' This principle is sometimes used to argue why certain constants of Nature (the ones in the Standard Model for instance) have the values they have. The problem with that is, in my opinion, the following.

The anthropic principle seems to work reasonably well to explain why certain constants are what they are, up to some degree of accuracy. It is probably true that if the fine structure constant $\alpha$ were only slightly different from the known value, 1/137.036 ... nuclear and atomic physics would give some substances quite different chemical properties and consequently also different abundances, such that human life would probably have become impossible. But would this also be true for the 24th decimal place of $\alpha$? And what about the 10 000th decimal place? Now I admit that physicists would have a hard time even to define this constant to such an accuracy, but it should be true that at least some phenomena in this universe do depend on such small details. But does the existence of humanity depend on all decimal places of $\alpha$? Probably not. So the anthropic principle only works up to a point.

In contrast, there is a version of the anthropic principle that is obviously correct: we are living on Earth, not Venus, Mars or the Moon. The reason is obvious: those places are (as yet) uninhabitable. Our domicile has a breathable atmosphere. The anthropic principle does explain why our atmosphere is breathable, and I guess we all accept this explanation. The difference between these two anthropic principles can easily be formulated in strictly mathematical terms. Stars and planets can be enumerated – that is, one could make a finite list of all of them in our present universe. The acceptable version of the anthropic principle asks us to choose one item of this denumerable list. The version that is not acceptable to me asks us to choose a physical constant out of a set of real numbers. These numbers are strictly innumerable. So I propose to accept the 'discrete anthropic principle' but to reject the 'continuous anthropic principle'.

The discrete anthropic principle states that there could be many Conway games, they form a discrete – denumerable – list. The principle allows us to pick the model we can live in as being our universe.

## 7 Perennial doubt

Thank you for allowing me to day-dream. The universal law does not – not yet? – exist. Until that time arrives we can just speculate about it. But, of course, we can also try to find out if there are no further compelling counter arguments.

One thing seems to be clear. Perhaps someone will come along with a proposal for a universal law. Perhaps this proposal will pass a certain number of tests. But all experiments have limited accuracy. How can we ever be sure that the theory is correct? Of course we can't. One could always keep the suspicion that the proposed law is nearly but not quite always and everywhere valid. Every now and then Nature could choose to deviate, either following a different law, or just following no law at all.

I would then argue that this is implausible. This is not Nature as we physicists came to know her. Nature's laws are basically simple, straightforward and absolute. There are no dispensations. You could call this 'professional experience' or 'intuition' or just 'faith', but completely convincing it will never be.

More serious is the fundamental denial of the necessity of any law at all. Many would argue that every now and then there will be 'divine intervention'. My personal belief is that 'God', or whoever it is out there, has ample opportunity to play tricks with the chaotic solutions of the equations so that suspension of the equations is not at all necessary if he (or she) wishes to divinely intervene.

The most difficult objection to counter is the denial that a physical law can be simple or defined at all at a local level and scale. It could be, as is indeed favored by many researchers of quantum theory, that there exists some untraceable action at a distance. There may exist laws that cover large objects or beings separately, even laws that give to people 'minds' and 'souls'. Laws could exist that allow for telepathic communications. And so on. If you believe any of this, you will have to reject our Theories of Everything, which leave no room whatsoever for such frivolities.

Physicists who search for their universal theory are no missionaries. We have not the least desire to convince anyone against his or her will into believing our theory. All we wish to do is marvel at Nature's beauty and simplicity. We have seen and tasted the beauty, simplicity and universality of our latest theories and models of the fundamental particles and the cosmos. We are now trying to uncover more of that. It is our belief that there is more. It is a challenge we cannot ignore. And whatever we find, there will be questions and challenges that will be left unanswered.

# 2 Theories of Everything

John D. Barrow

## Introduction

Despite the topicality of Theories of Everything[1] in the literature of science and its popular chronicles, they are at root a new edition of something very old indeed. If we cast our eyes over a range of ancient mythological accounts of the world we soon find that we have before us the first Theories of Everything. Their authors composed elaborate stories in which there was a place for everything and everything had its place.[2] These were not in any modern sense scientific theories about the world, but tapestries within which the known and the unknown could be interwoven to produce a single meaningful picture in which the authors could place themselves with a confidence born of their interpretation of the world around them. In time, as more things were discovered and added to the stories, so they became increasingly contrived and complicated. Moreover, whilst these accounts aimed at great breadth when assimilating perceived truths about the world into a single coherent whole, they were totally lacking in depth. That is, in the ability to extract more from their story than what was put into it in the first place. Modern scientific theories about the world place great emphasis upon depth – the ability to predict new things and explain phenomena not incorporated in the specification of the theory initially. For example, an explanation for the living world that maintained it to have been created ready-made just hundreds of years ago but accompanied by a fossil record with the appearance of billions of years of antiquity, certainly has breadth; but it is shallow. Experience teaches us that it is most efficient to begin with a theory that is narrow but deep and then seek to extend it into a description that is both broad and deep.[1] If we begin with a picture that is broad and shallow we have insufficient guidance to graduate to a correct description that is broad and deep.

In recent years there has been renewed interest in the possibility of a Theory of Everything.[1] In what follows we shall see what is meant by a Theory of Everything and how, while it may be necessary for our description of the Universe and its contents, it is far from sufficient to complete that understanding. We cannot 'reduce' everything we see to a Theory of Everything of the particle physicists' sort. Other factors must enter to complete the scientific description of the Universe. One of the lessons that will emerge from our account is the extent to which it is dangerous to draw conclusions about 'science' or the 'scientific method' in general in the course of discussing an issue like reductionism or the relative merits of religion and science. Local sciences, like biology or chemistry, are quite different to astronomy or particle physics. In the local sciences one can gather virtually any data one likes, perform any experiment and (most importantly of all) one has control over possible sources of bias introduced by the experimental set-up or the process of gathering observations. Experiments can be repeated in different ways. In astronomy this is no longer the case: we cannot experiment with the Universe: we just have to take what is on offer. What we see is inevitably biased by our existence and the means we use for seeing: intrinsically bright objects are over-represented in our surveys. Likewise, in high-energy particle physics there are serious limitations imposed upon our ability to experiment. We cannot achieve the very high energies that are required to unlock many of the secrets of the elementary particle world by direct experiment. The philosophy of science has said a lot about scientific method under the assumption of an ideal environment in which any desired experiment is possible. It has not, to my knowledge, addressed the reality of limited experimental possibility with the same enthusiasm.

## Order out of chaos

Suppose you encounter two sequences of digits. The first has the form

. . . 001001001001001001. . . .

whilst the second has the form

. . . 01001011010111101 0010 . . .

Now you are asked if these sequences are random or ordered. Clearly the first appears to be ordered and you say this because it is possible to 'see' a pattern in it; that is, we can replace the sequence by a rule that allows us to remember it or convey it to others without simply listing its contents. Thus we will call a sequence non-random if it can be abbreviated by a formula or a rule shorter than itself. If this is so we say that it is *compressible*.[3] On the other hand, if, as appears to be the case for the second sequence (which was generated by tossing a coin) there is no abbreviation or formula which can capture its information content then we say that it is *incompressible*. If we want to tell our friends about the incompressible sequence then we will have to list it in full. There is no encapsulation of its information content shorter than itself.

This simple idea allows us to draw some lessons about the scientific search for a Theory of Everything. We might define science to be the search for compressions. We observe the world in all possible ways and gather facts about it; but this is not science. We are not content like archivists simply to gather up a record of everything that has ever happened. Instead we look for patterns in those facts, compressions of the information on offer, and these patterns we have come to call the laws of Nature. The search for a Theory of Everything is the quest for an ultimate compression of the world. Interestingly, Chaitin's proof of Gödel's incompleteness theorem[4] using the concepts of complexity and compression reveals that Gödel's theorem is equivalent to the fact that one cannot prove a sequence to be incompressible. We can never prove a compression to be the ultimate one; there might still be a deeper and simpler unification waiting to be found.

There is a further point that we might raise regarding the quest for a Theory of Everything – if it exists. We might wonder whether such a theory is buried deep (perhaps infinitely deep) in the nature of the Universe or whether it lies rather shallow. One suspects it to lie deep in the structure of things and so it might appear a most anti-Copernican over-confidence to expect that we would be able to fathom it after just a few hundred years of serious study of the laws of Nature, aided by limited observations of the world by relatively few individuals. There appears to be no good evolutionary reason why our intellectual capabilities need be so great as to unravel the ultimate laws of Nature unless those ultimate laws are simply a vast elaboration of very simple principles, like counting or comparing,[5] which are employed in local laws. Of course, the unlikelihood of our success is no reason not to try. We just should not have unrealistic expectations about the chances of success and the magnitude of the task.

### Laws of Nature

Our discussion of the compressibility of sequences has taught us that pattern, or symmetry, is equivalent to laws or rules of change.[6] Classical laws of change, like Newton's law of momentum conservation, are equivalent to the invariance of some quantity or pattern. These equivalences only became known long after the formulation of the laws of motion governing allowed changes. This strikes a chord with the traditional Platonic tradition which places emphasis upon the unchanging, atemporal aspects of the world as the key to its fundamental structures. These timeless attributes, or 'forms' as Plato called them, seem to have emerged with the passage of time as the laws of Nature or the invariances and conserved quantities (like energy and momentum) of modern physics.

Since 1973 this focus upon symmetry has taken centre stage in the study of elementary particle physics and the laws governing the fundamental interactions of Nature. Symmetry is now taken as the primary guide into the structure of the elementary particle world, and the laws of change are derived from the requirement that particular symmetries, often of a highly abstract character, be preserved. Such theories are called 'gauge theories'.[7] The currently successful theories of four known forces of Nature – the electromagnetic, weak, strong and gravitational forces – are all gauge theories. These theories require the existence of the forces they govern in order to preserve the invariances upon which they are based. They are also able to dictate the character of the elementary particles of matter that they govern. In these respects they differ from the classical laws of Newton which, since they governed the motions of all particles, could say nothing about the properties of those particles. The reason for this added dimension is that the elementary particle world that the gauge theories rule, in contrast to the macroscopic world, is populated by collections of identical particles. ('Once you've seen one electron you've seen 'em all.')

The use of symmetry in this powerful way enables entire systems of natural laws to be derived from the requirement that a certain abstract pattern be invariant in the Universe. Subsequently, the predictions of this system of laws can be compared with the actual course of Nature. This is the opposite route to that which might have been followed a century ago. Then, the systematic study of events would have led to systems of mathematical equations giving the laws of change; afterwards, the fact that they are equivalent to some global or local invariance might be recognized.

The generation of theories for each of the separate interactions of Nature has motivated the search for a unification of those theories into more comprehensive editions. They are based upon larger symmetries within which the smaller symmetries respected by the individual forces of Nature might be accommodated in an interlocking fashion that places some new constraint upon their allowed forms. So far this strategy has resulted in a successful, experimentally tested, unification of the electromagnetic and weak interactions, and a number of purely theoretical proposals for a further unification with the strong interaction ('grand unification') and ultimately a four-fold unification with the gravitational force to produce a so-called Theory of Everything.[1] The pattern of unification that has occurred over the last three hundred years is shown in Fig. 1.

The current favoured candidate for a Theory of Everything is a superstring theory first developed by Michael Green and John Schwartz.[8] Elementary descriptions of its workings can be found elsewhere.[1,9] Suffice it to say that the enormous interest that these theories have attracted over the last nine years can be attributed to the fact that they revealed that the requirement of logical self-consistency – suspected of being a rather weak constraint upon a Theory of Everything – turned out to be enormously restrictive. At first it was believed

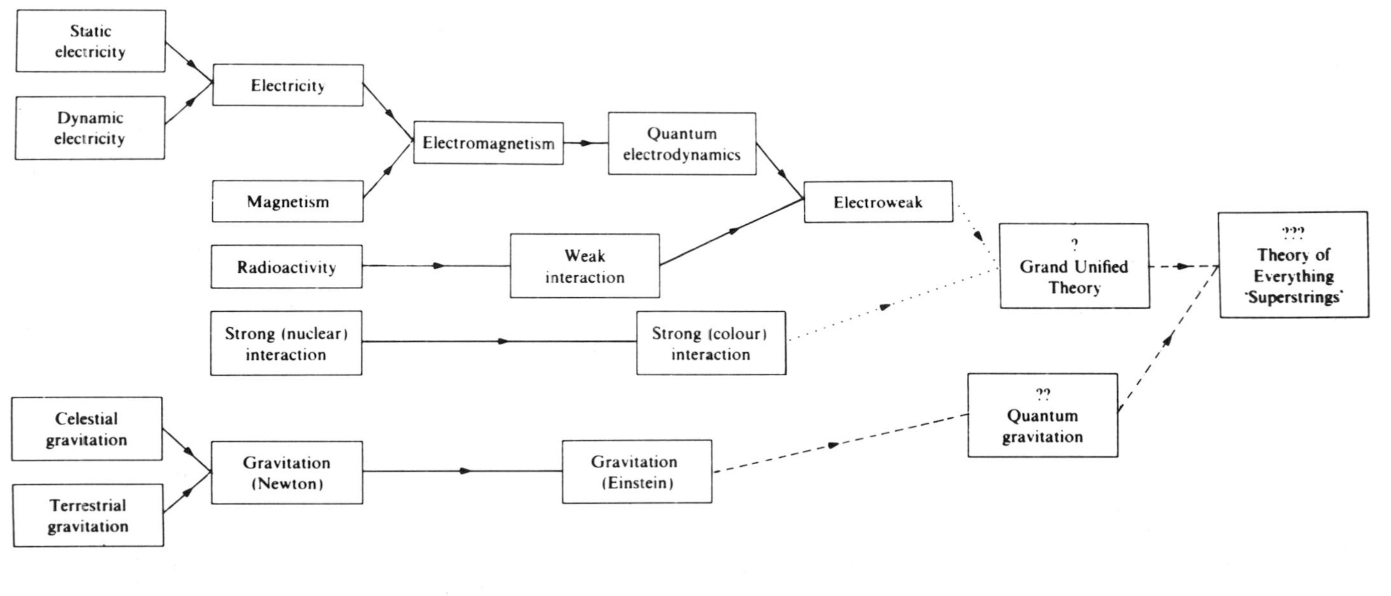


Fig. 1. The development of different theories of physics showing how theories of different forces of Nature have become unified and refined. The dotted lines represent unifications which have so far been made only theoretically and have yet to be confirmed by experimental evidence.

that it narrowed the alternative down to just two possible symmetries underlying the Theory of Everything. Subsequently, the situation has been found to be rather more complicated than first imagined and superstring theories have been found to require new types of mathematics for their elucidation.

The important lesson to be learned from this part of our discussion is that Theories of Everything, as currently conceived, are simply attempts to encapsulate all the laws governing the fundamental forces of Nature within a single law of Nature derived from the preservation of a single overarching symmetry. We might add that at present four fundamental forces are known, of which the weakest is gravitation. There might exist other far weaker forces of Nature that are too weak for us to detect (perhaps ever) but whose existence is necessary to fix the logical necessity of that single Theory of Everything.

### Outcomes and broken symmetries

If you were to engage particle physicists in a conversation about the nature of the world they might soon be regaling you with a story about how simple and symmetrical the world really is if only you look at things in the right way. But when you return to contemplate the real world you know that it is far from simple. For the psychologist, the economist, the botanist or the zoologist the world is complicated. It is a higgledy-piggledy of complex events whose nature owes more to their persistence or stability over time than any mysterious attraction for symmetry or simplicity. So who is right? Is the world really simple, as the physicist said, or is it as complicated as everyone else seems to think?[10]

The answer to this question reveals one of the deep subtleties of the Universe's structure. When we look around us we do not observe the laws of Nature; rather, we see the outcomes of those laws. There is a world of difference. Outcomes are much more complicated than the underlying laws because they do not have to respect the symmetries displayed by the laws. By this means it is possible to have a world which displays complicated asymmetrical structures (like ourselves) and yet is governed by very simple, symmetrical laws. Consider the following simple example. Suppose I balance a ball at the apex of a cone. If I were to release the ball then the law of gravitation will determine its subsequent motion. But gravity has no preference for any particular direction in the Universe; it is entirely democratic in that respect. Yet when I release the ball it will always fall in some particular direction, either because it was given a little push in one direction or as a result of quantum fluctuations which do not permit an unstable equilibrium state to persist. Thus, in the outcome of the ball falling down, the directional symmetry of the law of gravity is broken. Take another example. You and I are at this moment situated at particular places in the Universe despite the fact that the laws of Nature display no preference for any one place in the Universe over any other. We are both (very complicated) outcomes of the laws of Nature which break their underlying symmetries with respect to positions in space. This teaches us why science is often so difficult. When we observe the world we see only the broken symmetries manifested through the *outcomes* of the laws of Nature and from them we must work backwards to unmask the hidden symmetries which characterize the laws behind the appearances.

We can now understand the answers that we obtained from the different scientists. Particle physicists work closest to the laws of Nature themselves and so are impressed by their simplicity and symmetry. That is the basis for their assertion about the simplicity of Nature. But biologists or meteorologists are occupied with the study of the complex outcomes of the laws rather than with the laws themselves. As a result they are more impressed by the complexities of Nature rather than by its laws. This dichotomy is displayed in Fig. 2.

The left-hand column represents the development of the Platonic perspective on the world with its emphasis upon the unchanging elements behind things – laws, conserved quantities, symmetries – whereas the right-hand column, with its stress upon time and change and the concatenation of complex happenings, is the fulfilment of the Aristotelian approach to understanding of the world. Until rather recently, physicists have focussed almost exclusively upon the study of the laws rather than the complex outcomes. This is not surprising.

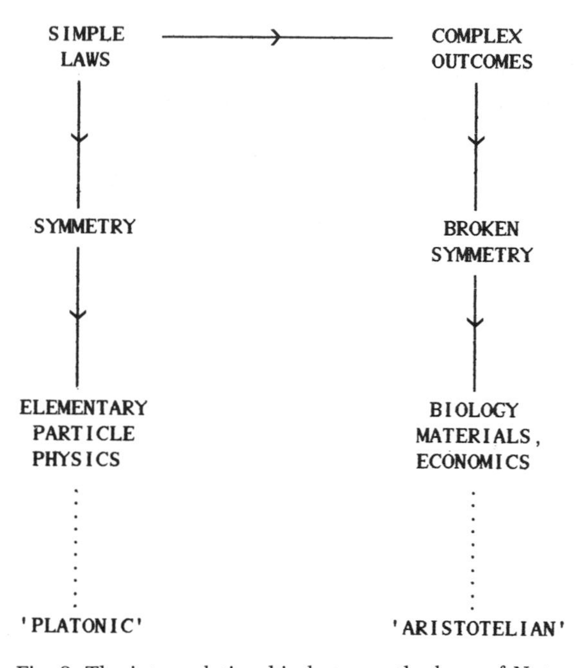


Fig. 2. The inter-relationship between the laws of Nature and the outcomes of those laws together with the scientific disciplines that focus primarily upon them. The left-hand thread from the laws of Nature is in the Platonic tradition whilst the right-hand line focusses upon temporal development and complicated outcomes and is associated with the Aristotelian perspective upon the world.

For the study of the outcomes is a far more difficult problem that requires the existence of powerful interactive computers with good graphics for its full implementation. It is no coincidence that the study of complexity and chaos[11] in that world of outcomes has gone hand in hand with the growing power and availability of low cost personal computers.

We see that the structure of the world around us cannot be explained by the laws of Nature alone. The broken symmetries around us may not allow us to deduce the underlying laws and a knowledge of those laws may not allow us to deduce the permitted outcomes. Indeed, the latter state of affairs is not uncommon in fundamental physics and is displayed in the current state of superstring theories. Theoretical physicists believe they have the laws (that is, the mathematical equations) but they are unable to deduce the outcomes of those laws (that is, find the solutions to the equations). Thus we see that whilst Theories of Everything may be necessary to understand the world we see around us, they are far from sufficient.

Of those complex outcomes of the laws of Nature much the most interesting are those that display what has become known as 'organized complexity'. A selection of these are shown in Fig. 3 in terms of their size (gauged by information storage capacity) versus their ability to process information (change one list of numbers into another list).

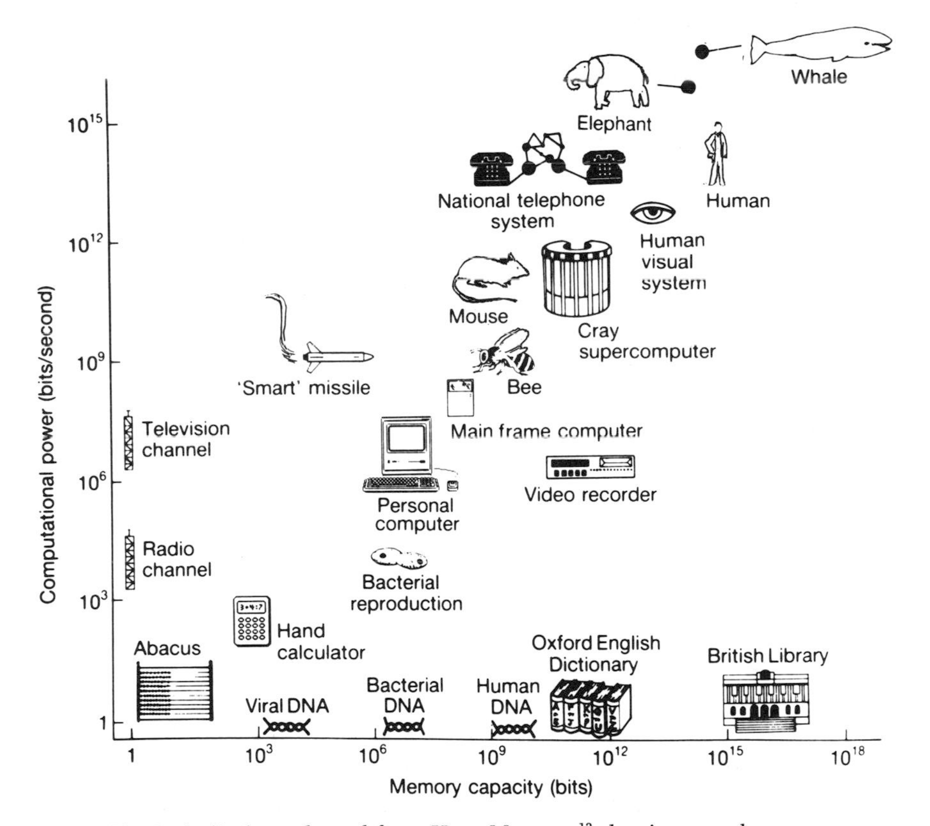


Fig. 3. A display, adapted from Hans Moravec,[12] showing complex organized structures in terms of their volume (in terms of capacity for storing information) and processing power (the speed at which they can change that information). We recognize the most complex structures as those at the top right of the graph.

Increasingly complex entities arise as we procede up the diagonal where increasing information storage capability grows hand-in-hand with the ability to transform that information into new forms. These complex systems are typified by the presence of feedback, self-organization and locally-purposeful behaviour. There might be no limit to the complexity of the entities that can exist farther and farther up the diagonal. Thus, for example, a complex phenomenon like high-temperature superconductivity,[13] which relies upon a very particular mixture of materials being brought together under special conditions, might never have been manifested in the Universe before the right mixtures were made on Earth in 1987. It is most unlikely that these mixtures occur naturally in the Universe and so that variety of complexity called 'high-temperature superconductivity' relies upon that other variety of complexity called 'intelligence' to act as a catalyst for its creation. Moreover, we might speculate that there exist new types of 'law' or 'principle' which govern the existence and evolution of complexity defined in some abstract sense.[14] These rules might be quite different from the laws of the particle physicist and not be based upon symmetry and invariance but upon principles of logic and information processing.

The defining characteristic of the structures in Fig. 3 is that they are more than the sum of their parts. They are what they are, they display the behaviour that they do, not because they are made of atoms or molecules, but because of the way in which their constituents are organized.[15] It is the circuit diagram of a neural network that is responsible for the complexity of its behaviour. The laws of electromagnetism alone are insufficient to explain the workings of a brain. We need to know how it is wired up. No Theory of Everything that the particle physicists supply us with will shed any light upon the workings of the human brain or the nervous system of an elephant.

So far we have discussed the outcomes of the laws of Nature using rather straightforward examples but we shall find that there are some aspects of the Universe that were once treated as unchanging parts of its constitution that have gradually begun to appear more and more like asymmetrical outcomes; indeed, the entire Universe may be one such asymmetrical outcome of an underlying law whose ultimate symmetry is hidden from us.

## Cosmology

The greatest discovery of twentieth-century science is that the Universe is expanding; that is, the distant clusters of galaxies are receding from each other at a velocity that increases linearly with their distance apart.[16] As a result the Universe has a history. The average cosmic environment is continuously cooling and rarifying as the Universe expands and ages. If we reverse the sense of expansion in our mind's eye and trace the Universe backwards in time we encounter epochs at which the Universe was too hot for stars, molecules or even atoms and nuclei to exist. Today, the Universe appears to have been expanding for about fifteen billion years. In Fig. 4 the general evolution of the Universe is mapped out.

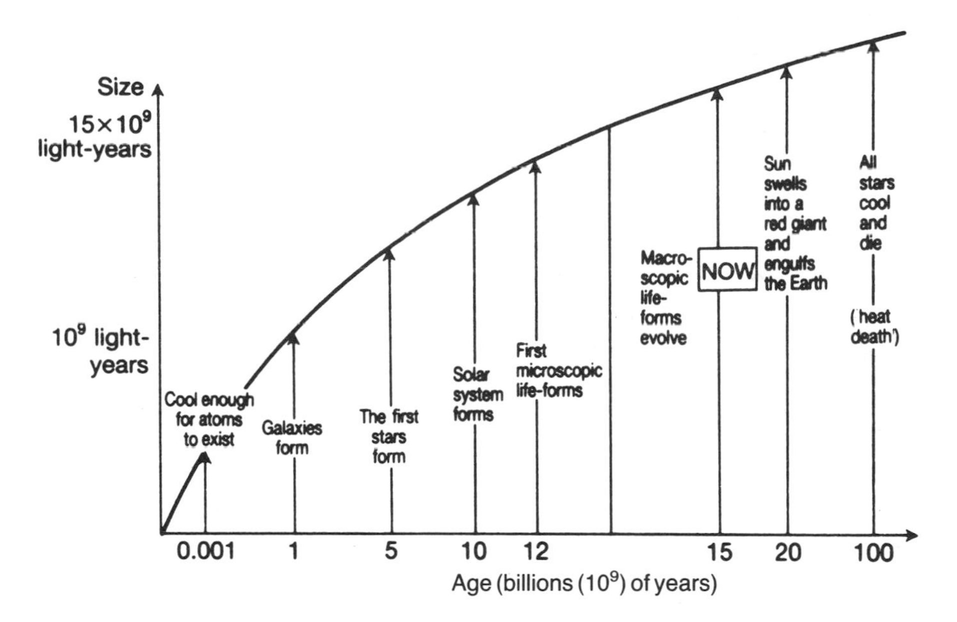


Fig. 4. The overall pattern of cosmic history from the time when the Universe was about a million years old until the far future. The expansion ensures that the ambient temperature of the Universe is steadily falling. Only after a particular period of time are conditions cool enough for the formation of atoms, molecules, and living complexity. In the future we expect an epoch when all the stars will have cooled and died. Thus there exists a niche of cosmic history before which life cannot evolve and after which conditions do not exist for its spontaneous evolution and persistence.

Biologists believe that *spontaneously* evolved life must be carbon based.[17] Given the prior existence of carbon-based life then all manner of other life-forms are undoubtedly possible – we appear to be aiding the evolution of 'artificial' silicon-based life-forms at present, although they exploit the subtleties of silicon physics rather than silicon chemistry as science fiction writers used to speculate about. Now carbon, and all the other biologically interesting elements heavier than helium, originate from nuclear reactions in the interiors of stars after about ten billion years of nuclear evolution. Our knowledge of the evolution of stars also teaches us that stable stars like the Sun will eventually exhaust their reserves of nuclear fuel and undergo a dramatic explosive phase of evolution before finally 'dying', unable to generate heat and energy. Hence there exists a narrow niche of cosmic history in which the spontaneous evolution of life must occur if it is ever to occur. At early times the building blocks necessary for complexity do not exist; at later times there are no stable hydrogen-burning stars to provide a steady environment. Obviously, we find ourselves living at a time in the history of the Universe that lies within this niche of life. This is a necessary precondition for our own existence.

The lesson that we should learn from this discussion is that the large-scale structure of the Universe is connected to the existence of living observers within it.[17] We can dramatize this by posing the question: why is the Universe so big? The visual horizon of the Universe is fifteen billion ($15 \times 10^9$) light years away. Within the volume that it encompasses there are about one hundred billion galaxies like the Milky Way, each containing about a hundred billion stars like the Sun and much else besides. Do we really need so much Universe? Could we not make do with a smaller economy-sized version?

We have seen that in order for the building blocks of any form of organised complexity to be present in the Universe we require at least ten billion years for the stars to produce and disperse elements heavier than helium throughout the Universe. So, because the Universe is expanding, it must be at least ten billion light years in size in order than it can contain 'observers' like ourselves. Thus the large size of the Universe should not surprise us: it is a necessary prerequisite for our own existence. Were the Universe to be just the size of the Milky Way galaxy, containing one hundred billion suns, then it would have been expanding for little more than a month – barely enough time to settle your credit card bill, let alone evolve complex observers.

This discussion highlights an important realization by cosmologists. They have recognized that 'observers' can only exist at particular times during the evolution of the Universe and, indeed, only in universes that grow old enough and big enough to produce biologically useful elements (that is, those heavier than helium). One can take this consideration a little further and add that observers can only exist in particular places in the Universe where conditions are cool enough and stable enough for their evolution and persistence. These considerations are of importance if there exists any random factor in the initial make-up or early evolution of the Universe which creates conditions that differ significantly from place to place. If so, then we need to take into account that our view of the Universe is biased by the fact that we necessarily observe it from one of those regions which satisfy the necessary conditions for the evolution of complexity.

One example of this state of affairs is provided by the most general edition of the inflationary universe theory.[18] This is an elaboration of the standard big bang picture of the expanding universe which proposes that during the very early stages of the expansion the expansion *accelerated* for a brief period. The standard (non-inflationary) big bang model manifests expansion that decelerates at all times regardless of whether the expansion continues forever or collapses to a big crunch in the future. This brief period of inflation has a variety of interesting consequences: it can guarantee that the visible universe possesses many of the mysterious properties that are observed. This accelerated phase will arise if certain types of matter exist during the very hot early stages of the Universe. If this matter (which we call the 'inflaton field') moves slowly enough then it drives a period of inflationary expansion. At the very early time when this is expected to occur light will have had time to travel only about $10^{-25}$ cm since the expansion began. Thus within each spherical region of this diameter there should exist a smooth, correlated environment; but conditions could be quite different over more widely separated dimensions because there will not have been enough time for signals to travel between different regions. The situation is represented in Fig. 5.

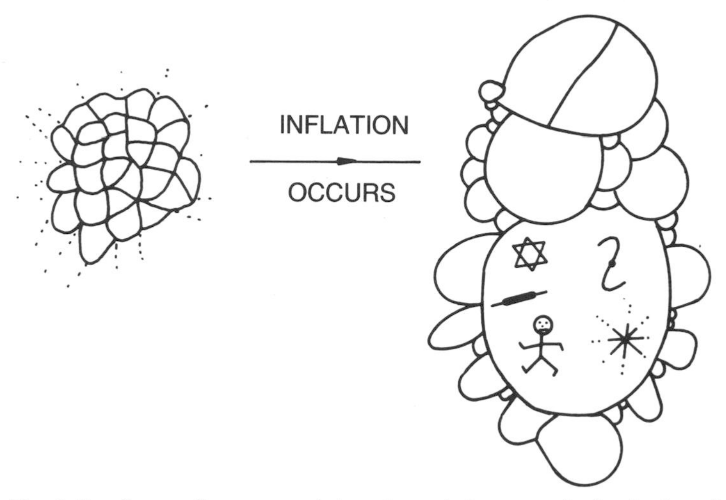


Fig. 5. Small causally-connected domains existing near the beginning of the expanding universe inflate separately by different amounts. We must find ourselves in one of those domains that expands to more than about ten billion light years in size. Only these large, old domains have enough space and time for the evolution of stars and the ensuing production of the elements heavier than helium which are necessary for the evolution of complex structures.

In each of those causally-disjoint regions the inflation field will evolve differently, reflecting the differing conditions and starting states within different regions. As a result, different regions will accelerate, or 'inflate' by different amounts in some random fashion. Some regions will inflate by large amounts and give rise to regions that remain expanding for more than ten billion years. Only in these regions will the formation and evolution of stars, of heavy elements and organized complexity be possible. We necessarily live in one of those regions.

## The visible universe

We should draw a distinction between '*The Universe*' which might be infinite in extent and the '*visible universe*' – that part of the Universe from which we have had time to receive light signals since the beginning of the expansion about fifteen billion years ago. The visible universe is finite and steadily getting bigger as the Universe ages. All our observations of the Universe are confined to the subset of the whole that is the visible part. Now take the region constituting the visible universe today, reverse the expansion of the Universe and determine the size of that region which, at any given time in the past, is going to expand to become our visible universe today.[1] When the Universe was about $10^{-35}$ s old, the present visible universe was contained within a sphere less than 1 cm in radius, as shown in Fig. 6.

This situation has a number of important consequences for any quest for a Theory of Everything. In particular, as can be seen from Fig. 7, our visible universe is the expanded image of an infinitesimal part of the entire expanse of initial conditions.

We can never know what the entire initial data space looks like. Moreover, whilst it is fashionable amongst cosmologists to attempt to formulate grand principles which specify the initial conditions (for example Roger Penrose's minimum Weyl entropy condition,[19] Alex Vilenkin's 'outgoing wave condition'[20] or James Hartle's and Stephen Hawking's 'no boundary condition'[21]), these principles specify averaged conditions over the entire initial data space. Ultimately, these conditions will be quantum statistical in character. However, such principles could never be tested even with all the information available in the visible universe. The entire visible universe is determined by a tiny part of the initial data space, which may well be atypical in certain respects in order that it satisfy the conditions necessary for the subsequent evolution of observers. Global principles about the initial state of the Universe, even if correct, may be of little use in understanding the structure of the visible part of the Universe because it evolves from an idiosyncratic part of an infinite span of initial conditions.

## Constants of Nature

The constants of physics – quantities like the mass of an electron or its electric charge – have traditionally been regarded as unchanging attributes of the Universe. An acid test of any Theory of Everything would be its ability to predict the value of these constants.[22]

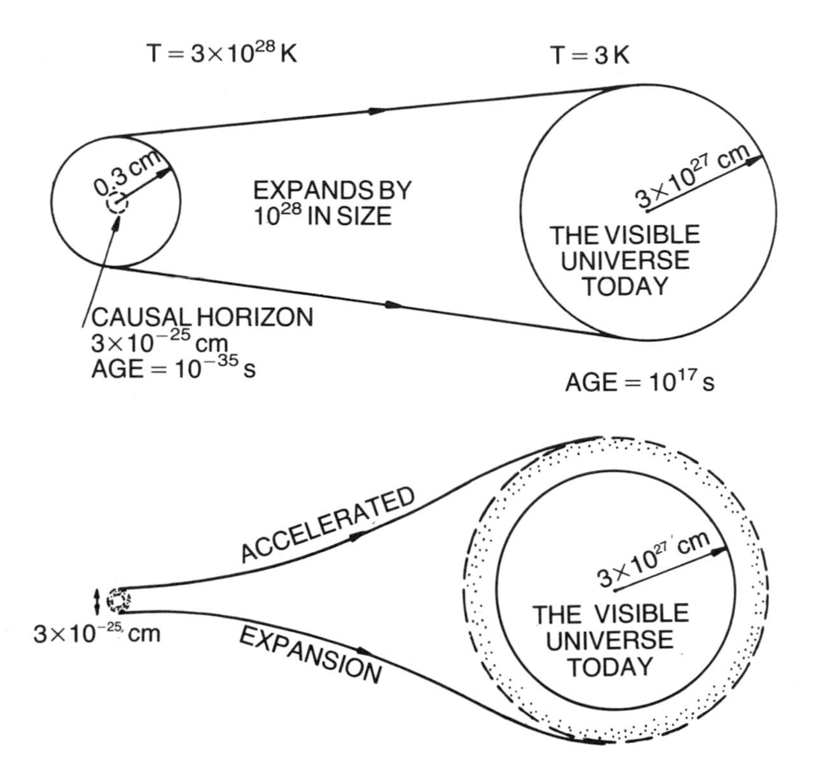


Fig. 6. The past history of the region defining our visible universe today. When the Universe was $10^{-35}$ s old this was 0.3 cm in radius. By contrast, the visible universe at that time was $10^{-25}$ cm in radius. Thus our visible universe evolved from a vast number of causally disjoint regions. Why then does it look so uniform from place to place over large scales? Why does the background radiation field have the same temperature in every direction to within a few parts in one hundred thousand? The inflationary universe picture can resolve this dilemma by accelerating the early expansion of the Universe for a brief period. This more rapid expansion enables the visible universe today to have expanded from a region of radius only $10^{-25}$cm, or smaller, in radius at the time of $10^{-35}$ s. This particular time is used for illustrative purposes. Inflation does not have to occur at this particular time although it could well have done so.

The values of the constants of Nature are what endow the Universe with its coarse-grained structure. The sizes of stars and planets, and to some extent also people, are determined by the values of these constants.[23] In addition, it has gradually been appreciated that the

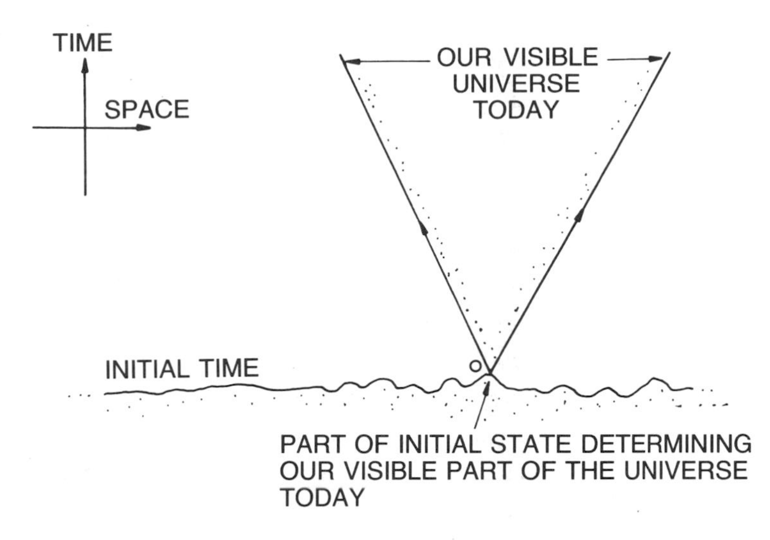


Fig. 7. A space and time diagram showing the development of our visible universe from its initial conditions. The structure of our entire visible universe is determined by conditions at the point O. These conditions may need to be atypical in order to permit the subsequent evolution of observers, and their detailed local form will not be determined by any global principle which determines the initial conditions quantum statistically.

necessary conditions for the evolution of organized complexity in the Universe are dependent upon a large number of remarkable 'coincidences' between the values of different constants. Were the constants governing electromagnetic or nuclear interactions to be very slightly altered in value then the chain of nuclear reactions that produce carbon in the Universe would fail to do so; change them a little more and neither atoms nor nuclei nor stable stars could exist.[17,23,24,25,26] So we believe that only those universes in which the constants of Nature lie in a very narrow range (including, of course, the actual values) can give rise to observers of any sort.

In recent years the attempts to develop a quantum cosmological model have uncovered the remarkable possibility that the constants of Nature may be predictable but their values will be determined quantum statistically by the space–time fabric of the Universe.[27]

If we relinquish the idea that the topological structure of the Universe is a smooth ball, then we picture it as a crenellated structure with a

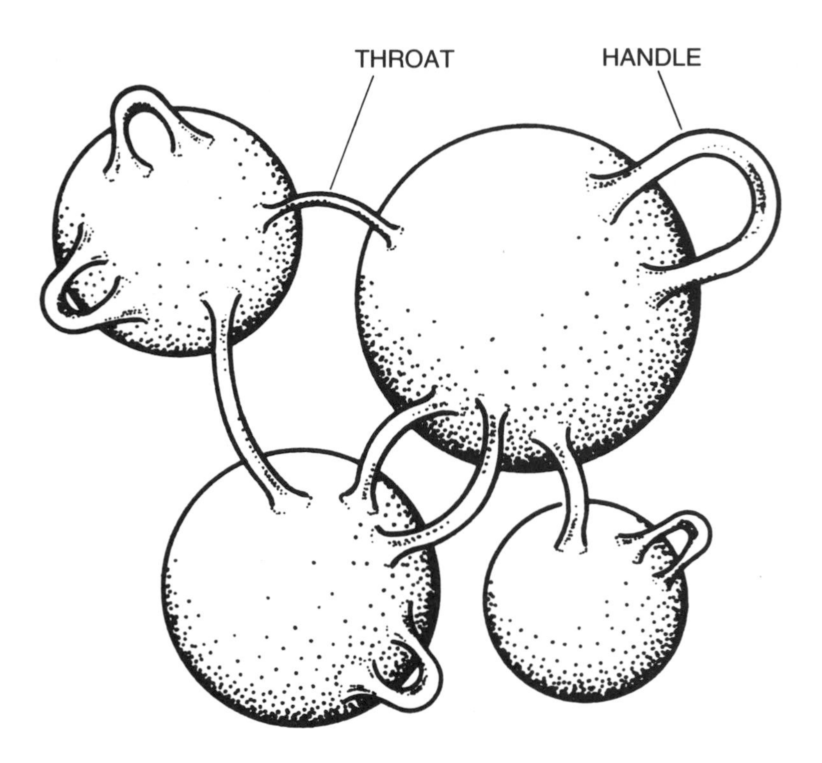


Fig. 8. Wormholes connecting different large sub-regions of the Universe to themselves by wormhole 'handles' and to other such regions by wormhole 'throats'.

network of wormholes connecting it to itself and to other extended regions of space–time as pictured in Fig. 8. The size of the throats of the wormholes is of order $10^{-33}$ cm. Now this turns out to be something more interesting than simply generalization for its own sake. It appears that the values of the observed constants of Nature on each of the large regions of space–time may be determined quantum statistically by the network of wormhole connections. Thus, even if those constants had their values determined initially by the logical strait-jacket of some unique Theory of Everything, those values would be shifted so that they would be observed today to have values given by a calculable quantum probability distribution. However, there is a subtlety here.[1] We might think that it is the business of science to compare the most probable value predicted with that observed. But why should the observed universe display the most probable value in any sense of the word 'probable'? We know that observers can only evolve if the constants lie in a very narrow range, so the situation might be that shown in Fig. 9.

We are interested, not in the most probable value of a constant, but only in the conditional probability of the constant taking a value that subsequently permits living observers to evolve. This may be very different to the unconditioned 'most probable' value.

This type of argument can be extended to consider the dimensionality of the part of space that becomes large. Superstring theories naturally predict that the Universe possesses more than three dimensions of space,[8,9] but it is expected that only some of these dimensions will expand to become very large. The rest must remain confined and imperceptibly small. Now we might ask why it should be *three* dimensions that grow large. Is this the only logical possibility or just a random

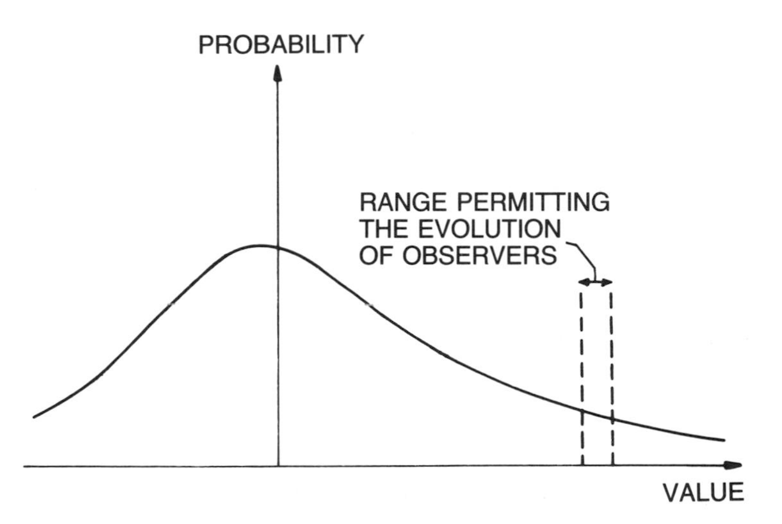


Fig. 9. A hypothetical prediction for the measured probability of a constant of Nature when the Universe is large and old (i.e. now) but the small range of values for which living complexity can exist is also shown. The latter range may be far from the most probable value predicted by the wormhole fluctuations. We should only be interested in the most probable value given that observers can subsequently evolve.

symmetry breaking that could have fallen out in another way, or perhaps has fallen out in different ways all over the Universe, so that regions beyond our horizon may have four or more large dimensions. Again, we know that the dimensionality of space is critically linked to the likelihood of observers evolving within it[28] (for instance, there are no atoms or stable gravitationally bound orbits in more than three dimensions). Similar considerations to those we have just introduced with regard to the probabilistic prediction of the constants apply to any random symmetry-breaking process that determines the number of space dimensions that become large.

These examples reiterate that the output of a Theory of Everything is not immediately comparable with the observed universe because, if there exists any random element in the initial structure of the Universe (and quantum theory makes this inevitable), we observe aspects of things that may not be typical. They are conditioned by the necessary conditions for the evolution of complexity. The observed universe should be thought of as being an outcome of the laws of Nature because we only observe part of what might be an infinite whole. Whilst a Theory of Everything is necessary to understand the observed part of the Universe, it is far from sufficient. We need to understand the role and nature of initial conditions; how our visible universe may be atypical in order to satisfy the necessary conditions for the evolution of observers; and how more complex symmetries break to allow the evolution of organized complexity in the Universe.

### What are the ultimate rules of the game?

Physicists tend to believe that a Theory of Everything will be some set of equations governing entities like points or strings that are equivalent to the preservation of some symmetry underlying things. This is an extrapolation of the direction in which particle physics has been moving for some time. A key assumption of such a picture is that it regards the laws of physics as being the bottom line and these laws govern a world of point particles or strings (or other exotica) that is a continuum. Another possibility is that the Universe is not, at root, a great symmetry, but a computation. The ultimate laws of Nature may be akin to software running on the hardware provided by elementary particles and energy. The laws of physics might then be derived from some more basic principles governing computation and logic. This view might have radical consequences for our appreciation of the subtlety of Nature because it seems to require that the world is discontinuous, like a computation, rather than a continuum. This makes the Universe a much more complicated place. If we count the number of discontinuous changes that can exist we find that there are infinitely many more of them than there are continuous changes. By regarding the bedrock structure of the Universe as a continuum we may not just be making a simplification but an infinite simplification.

## Notes


1. J. D. Barrow, *Theories of Everything: the quest for ultimate explanation* (Oxford University Press, 1991 and Vintage pbk, 1992).
2. C. H. Long, *Alpha: the myths of creation* (G. Braziller, 1963).
3. G. Chaitin, *Algorithmic Information Theory* (Cambridge University Press, 1987).
4. G. Chaitin, 'Randomness in arithmetic', *Scientific American* (July 1980), 80.
5. J. D. Barrow, *Pi in the Sky: Counting, Thinking and Being* (Oxford University Press, 1992).
6. J. D. Barrow, *The World Within the World* (Oxford University Press, 1988). R. Feynman, *The Character of Physical Law* (MIT Press, 1965).
7. H. Pagels, *Perfect Symmetry* (Michael Joseph, 1985). A. Zee, *Fearful Symmetry: the search for beauty in modern physics* (Macmillan, 1986). S. Weinberg, *The Discovery of Sub-atomic Particles* (W. H. Freeman, 1983).
8. M. Green, J. Schwartz and E. Witten, *Superstring Theory*, (two vols, Cambridge University Press, 1987).
9. M. Green, 'Superstrings', *Scientific American* (September 1986), 48. D. Bailin, 'Why Superstrings?', *Contemporary Physics* 30 (1989), p. 237, P. C. W. Davies and J. R. Brown (eds.), *Superstrings: A Theory of Everything?* (Cambridge University Press, 1988).
10. J. D. Barrow, 'Platonic relationships in the Universe', *New Scientist* (20 April 1991), p. 40.
11. J. Gleick, *Chaos: making a new science* (Viking, 1987), I. Stewart, *Does God Play Dice: the mathematics of chaos* (Blackwell, Oxford, 1989). B. Mandelbrot, *The Fractal Geometry of Nature* (W. H. Freeman, 1982).
12. H. Moravec, *Mind Children* (Harvard University Press, 1988).
13. C. Gough, 'Challenges of High-$T_c$', *Physics World* (December 1991), p. 26.
14. S. Lloyd and H. Pagels, 'Complexity as thermodynamic depth', *Annals of Physics* (New York, 1988), 188, 186.

15. P. C. W. Davies, *The Cosmic Blueprint* (Heinemann, 1987).
16. S. Weinberg, *The First Three Minutes* (A. Deutsch, 1977). J. D. Barrow and J. Silk, *The Left Hand of Creation* (2nd edn. Oxford University Press, 1994).
17. J. D. Barrow and F. J. Tipler, *The Anthropic Cosmological Principle* (Oxford University Press, 1986).
18. A. Guth, 'The inflationary Universe: a possible solution to the horizon and flatness problems, *Physical Review D* 23, (1981), p. 347. A. Guth and P. Steinhardt, 'The inflationary Universe', *Scientific American* (May 1984), p. 116. J. D. Barrow, 'The Inflationary Universe: modern developments', *Quarterly Journal of the Royal Astronomical Society* 29 (1988), p. 101.
19. R. Penrose, *The Emperor's New Mind* (Oxford University Press, 1989).
20. A. Vilenkin, 'Boundary conditions in quantum cosmology', *Physical Review D* **33** (1982), 3560.
21. J. B. Hartle and S. W. Hawking, 'The wave function of the Universe', *Physical Review D* **28** (1983), 2960. S. W. Hawking, *A Brief History of Time* (Bantam, 1988).
22. An infamous, completely unsuccessful programme of this sort was Arthur Eddington's quest for a 'Fundamental Theory', see A. S. Eddington, *Fundamental Theory* (Cambridge University Press, 1946) for an edited posthumous presentation, and A. S. Eddington, *New Pathways in Science* (Cambridge University Press, 1935) for a more popular account of its early progress. See also Barrow and Tipler (note 17) for an analysis.
23. J. D. Barrow, 'The mysterious law of large numbers', in S. Bergia and B. Bertotti (eds.), *Modern Cosmology in Retrospect* (Cambridge University Press, 1990).
24. B. J. Carr and M. J. Rees, 'The Anthropic Principle and the structure of the physical world', *Nature* 278 (1979), 605.
25. P. C. W. Davies, *The Accidental Universe*, (Cambridge University Press, 1982).
26. J. Gribbin and M. J. Rees, *Cosmic Coincidences: Dark Matter, Mankind, and Anthropic Cosmology* (Bantam, 1989).
27. S. Coleman, 'Why there is Something rather than Nothing', *Nuclear Physics B* 310 (1988), 643. S. W. Hawking, 'Wormholes in space-time', *Physical Review D* **37**, (1988), 904. S. W. Hawking, 'Baby universes', *Modern Physics Letters A* 5, (1990), 453.
28. J. D. Barrow, 'Dimensionality', *Philosophical Transactions of the Royal Society of London A* 310 (1983), 337.

# 3 The scientific view of the world: introduction

Dennis Dieks

It is my aim to introduce the themes brought up by Paul Feyerabend, Ernan McMullin and Bas van Fraassen by means of a brief overview – partly historical and partly systematical – of the realist and empiricist positions with respect to science. My selection and presentation of topics has been influenced by the wish to make a connection with some of the points raised in other contributions to this volume, especially the earlier chapters. (What do the results of modern physics tell us about reality?)

## 1 The traditional view of science

In the traditional Aristotelian view, science is the unique enterprise in which humanity discovers fundamental truths about the world in a systematic, rational, way. According to this view the result of scientific investigation is *knowledge*, to be contrasted with mere opinion; knowledge is defined as provably true, fundamental and universal. There are a number of characteristic ingredients in this conception.

*Realism*
Inherent in the traditional view is the realist position that science deals with an independent external world, which for its greater part is not accessible to the human senses but can nevertheless be discovered, investigated and described. Science actually *focuses* on these unobservable traits of the world, in order to give fundamental explanations.

*Science is fundamental*
Science is considered to be able to derive explanations for a wide range of phenomena from a limited number of basic principles. These prin-

ciples are not located on the level of what is observable; they pertain to a *deeper* level of reality, the level of *causes* (or grounds) *behind* what we can hear, see and feel.

*There is a well-defined scientific method*

It is possible to obtain such fundamental knowledge due to the existence of a definite scientific method. It is essential that this method is *ampliative*: it leads from observable phenomena to hidden causes and essences. In other words, something is *added* to what is contained in observation alone. According to Aristotle such a method exists due to the fact that humans possess a special innate faculty, an immediate insight or 'intuition', by means of which they can recognize fundamental facts about the world once they encounter them. Although scientific analysis starts by making observations in Nature, and finding candidates for truth on the basis of those observations (everything that is in the intellect has come to it through the senses), the guarantee of having found truth can only be reached through the power of intuition. Thus, in their intuition humans have a special organ to look beneath the surface of the phenomena.

*Science is universal*

Science is considered to be universal in more than one respect. The scientific method is taken to apply everywhere where knowledge is possible; not only to natural processes, but also to the realm of values and norms – indeed, there is no clear watershed between physical science and normative considerations in Aristotelian science. Besides this, scientific truth is supposed to be universal in that it is valid for all times and places, and thus totally culture-independent. It follows that a scientific fact, once uncovered, can never be in need of revision.

*Monism*

Often (though not always) the conviction that science can penetrate to the deepest levels of reality has been combined with the idea that the 'deep layers' encountered in physical investigations are also fundamental to other sciences. This is motivated by the *monistic* conception that in the final analysis there is only one world, consisting of only one kind of 'stuff', which can be described completely by physics. If this is accepted, it becomes natural to think of scientific disciplines as hierarchically ordered, with physics on the most fundamental level. Even in antiquity such ideas led to the formulation of proposals for 'Theories of Everything' (by the Atomists, for example).

## 2 Empiricist protests

From the very beginning protests have been voiced against this view of science. In antiquity sceptics denied that humans, with their limited capabilities of observation and reasoning, could ever achieve certain knowledge of underlying causes. In the late Middle Ages the Nominalists continued this line by arguing that humans are incapable of 'looking behind the scenes' of natural phenomena. They linked this with a voluntarist theology: God is completely free in assigning (and changing, if He wishes) causes for observable states of affairs. It is therefore impossible to deduce the mechanisms *behind* what we can observe from observation itself. All we can see is what actually happens, and regularities in these actual events. Whether things happen by necessity (in other terms: according to a 'law'), instead of as a mere coincidental regularity, is outside of observation and cannot be ascertained by us. In fact, God could make the same things happen either as necessary events or as things He just willed on each separate occasion. It follows that there cannot be a scientific method that leads with certainty from observational results to hidden causes. In empirical science we should therefore confine ourselves to the observable surface of the world; we should not speak of underlying essences, necessities, causes and laws. Only theology can attempt to penetrate to this level.

After the Scientific Revolution arguments of this type gained force. This is because it was central in the ideology of the Scientific Revolution to abandon the Aristotelian idea that humans possess a special faculty to identify fundamental truths behind the observable; the position was officially adopted that all scientific knowledge should be derived from observation. But then sceptical arguments become very urgent. How can scientific knowledge be more than a catalogue of empirical results if observation is its sole source?

David Hume went farthest in this direction: he concluded that knowledge about underlying causes and about what is going to happen in the future is impossible. So-called scientific knowledge, to the extent that it transcends actual observation, lacks all justification. In order to make room for the kind of life we normally live, Hume proposed a pragmatic (and somewhat schizophrenic) attitude: we should act *as if* we possessed knowledge about 'causes' and 'laws', because this is socially expected from us. But in our private moments we should realize that we know nothing.

Later empiricists responded to the problem in a more positive fashion. According to Ernst Mach it is true that in principle science only gives us a summary of empirical results; it cannot give us certainty about the future. The traditional idea that science affords provable knowledge should therefore be abandoned. But science *can* give us *hypotheses* about what is going to happen; in other words, hypotheses about regularities on the level of the observable. Such hypotheses are good enough: they are very useful and efficient. They result from compressing the huge amount of empirical data, so that the simplest and most economical description is achieved – this is what the 'scientific method' consists in, according to Mach. Science is very important, also in daily life: by their efficiency, scientific descriptions provide us with a means to enhance our chances in the struggle for existence. Science thus is an *instrument* that helps us to form expectations and to make predictions.

From a viewpoint like Mach's, science can perhaps still be said to be fundamental and universal. But the meaning of these terms is now very different from what they signified for Aristotle. The *fundamentality* of Machian science consists in its simplicity of form, its wide scope and its associated utility. In other words, it is a pragmatic fundamentality. There is no ontological claim that there is a reality *behind* the phenomena which corresponds to the mathematical formalism used in scientific theories, a reality more basic than the phenomena themselves. The *universality* of science is in the universal applicability of its method: the method of collecting empirical data and compressing them by means of the formulation of simple regularities and theoretical relations (with the status of hypotheses). According to Mach, the scientific attitude is first of all a critical one: empirical data always have the last word and there is no place for dogmas, sacrosanct theories and *a priori* statements. These principles clearly define the limitations of the scientific enterprise. Science can only give descriptions (and predictions) of factual states of affairs; norms and values cannot be founded by science and are outside its scope. Moreover, we can never be sure that what we accept today will still be acceptable tomorrow: scientific theories are hypothetical, cannot be proved and are likely to be rejected in the future.

Summing up, the aim of science is not the search for timeless Truth about the essential aspects of reality, but the search for the most useful, general and economical scheme (up to now) that 'saves the phenomena'. This point of view has become known as 'instrumentalism'.

### 2.1 Pluralism

Mach was a monist in that he maintained that elementary building blocks of observation, 'elements of sensation' or 'sense data', constituted the basis for all scientific disciplines. But we also encounter a pluralistic theme: Mach stressed that different perspectives are possible in scientific investigations. The results obtained from those different viewpoints can give rise to different scientific disciplines which possess relative autonomy.

We can, for example, apply the scientific method to phenomena outside of us; this eventually leads to physics. If we direct our attention to phenomena inside ourselves, this leads to psychology; in analogous fashion physiology, sociology, etc., arise. In each area regularities can be found. These stand beside each other, as independent results within their own territory. Of course, there may be correlations between them. Thus, correlations exist between on the one hand the psychological regularities found when we direct our attention to the relations between thoughts and ideas and on the other hand the physiological regularities in the relations between bodily reactions. But that does not imply that physiology is reducible to psychology, or vice versa. The study of the relations between psychological phenomena and physiology requires a *third* point of view, transcending the perspectives of psychology or physiology taken separately. In the context of this broader viewpoint psychology and physiology are on the same footing.

Within an empiricist outlook like Mach's it is therefore not self-evident that one discipline should be considered more fundamental than the others. The basic reason behind this is that scientific disciplines are not considered as attempts to describe one hidden deeper reality.

The logical empiricists of the twentieth century took a great deal of their inspiration from Mach (although many of them came to their empiricist position via Neo-Kantian philosophy). One of their main goals was to make the empiricist ideas precise by elaborating them in a formal way. To this end, the logical empiricists made logical analyses of scientific theories and tried to show exactly how these theories are basically only schemes for ordering observational data. They also put much effort into analyzing 'reduction relations' between theories, for instance between thermodynamics and statistical mechanics. What is involved here are interrelations between *theories*, as opposed to an ordering of different levels of reality. From a strictly empiricist point of view we are dealing with relations between observable phenomena as they are ordered by the various theories in question; there need not be an implication that one theory pertains to a deeper level of reality than another. The sense of fundamentality which is at stake here is first of all that of generality, that one theory is more comprehensive than another.

## 3 Modern scientific realism

I think it is fair to say that the exponents of the empiricist and sceptical traditions just sketched have always constituted a minority of dissenters. Also after the Scientific Revolution, the traditional notion that science attempts to discover the truth about the deepest level of reality, and is actually able to penetrate to this level, has been acclaimed by the majority of scientists and philosophers. Witnesses to this are, for example, Newton's optimism about the possibilities of the experimental, inductive, method to discover unchanging truths, also about unobservable things, and Einstein's conviction that the human intellect is able to grasp the (mathematical) essence of reality. In recent years realism has received an additional impetus among philosophers of science from the failure of the logical-empiricist analyses of science.

In order to make clear how in principle scientific theories only order the results of observation, the logical positivists had introduced a distinction between two classes of scientific terms. They maintained that each term falls into one of these two classes. 'Observation terms' do not involve theory and refer to directly observable things. 'Theoretical terms' (such as atom, molecule and quark) get their meaning via relations, determined by the theory, to observation terms. The idea was that it should be possible to translate all expressions containing theoretical language into expressions of the unproblematic, hygienic, observation language. This would indeed demonstrate the possibility of principle to formulate a theory entirely in terms of observable states of affairs.

But the distinction between 'observation terms' and 'theoretical terms', in the way the logical empiricists envisaged it, has turned out to be untenable. A chief reason is that *all* scientific terms depend for their meaning on a theoretical background. Even everyday terms referring to clearly observable things are 'theory-laden', as many philosophers have urged during the last few decades. All 'seeing' is seeing *that something is the case*; the use of concepts, which is indispensable in reporting the results of observation, necessarily introduces theoretical elements. There are no theory-free observation terms, and there is therefore no linguistic difference of principle between terms referring to everyday objects and terms like 'atom' or 'quark'. According to the realist this argument shows that all terms in a scientific theory possess the same status, and therefore that all such terms must be considered as equally descriptive of reality. Atoms and quarks must consequently be assumed to exist in the same way as tables and chairs. It is the aim of scientific theories to describe atoms and quarks just as precisely as we can describe everyday objects.

### 3.1 Realist strategies

The examples of Newton and Einstein mentioned above already show that unanimity about the *method* to be followed in order to achieve the aims of science no longer exists. The conviction that there is a collection of rules whose application will automatically result in the discovery of scientific truths has practically disappeared among scientists. Instead, the view has gained wide spread acceptance that scientific progress is a process of trial and error, in which *choices* have often to be made between alternatives which are equally compatible with the empirical data. Although such choices may be rationally justified (with the aid of such criteria as simplicity, 'naturalness', avoidance of the *ad hoc*, explanatory power), they are not forced upon us by logic or by a unique Scientific Method. The resulting laws and theories are hypothetical; one can have reasons to expect or hope that they are true about reality, but there is no automatic guarantee for this.

If scientists themselves do not believe that we possess a special intuition that makes it possible to perceive reality directly as it is, or that there is a special set of rules with which we can deduce the true nature of reality from phenomena, the question arises why we should believe any particular picture offered to us as a description of reality. An important realistic stratagem here is to say that scientific explanations can only be acceptable – can only be considered as really explaining – if we are convinced of the actual existence of the unobservable entities the explanations refer to. It is therefore rationally compelling to believe in the (approximate) truth of the best explanation, the one offered by the scientific theory we accept as the best. In other words, acceptance of a theory as explanatory implies belief in its (approximate) truth. In addition, realists sometimes hold that the success of science itself needs explanation. According to them, the best explanation is that there is an ever better fit between successive scientific theories and the real world. Arguing in the same way as with the explanations provided by science itself, but now at a meta-level, this leads to the conclusion that the most rational position to adopt is that our scientific theories indeed furnish ever better descriptions of reality.

What we have here are examples of an ampliative scheme of argument ('retroduction') known as 'inference to the best explanation'.

### 3.2 The modern realist outlook

Although ideas about the scientific method have very much changed, the general outlook of scientific realism has stayed close to traditional conceptions. The realist maintains that the aim of science is to discover and describe the outside world that is there independently of our (contingent) human presence. It is not the most important task of science to describe observable phenomena, but to *explain* them. Fundamental explanations should take their starting-point on the most fundamental level of reality, hidden behind the phenomena.

Concerning norms and values, it has gradually become a commonplace that science is situated on the 'is-side' of the is/ought distinction. Science depicts the structure of the world, but does not discover norms and values. In this respect important elements of the classical ideal of science and the scientific world-picture have been left behind. Nevertheless, the ideas that science tells us how the world really is, and that the scientific picture of the world is the most fundamental one that can bc had, can and do influence norms. Actually, these realistic ideas play a substantial role in practice. Among scientists, and to some extent among the public at large, there *is* a (perhaps somewhat vague) scientific world-picture; and this picture has a considerable influence on attitudes with respect to our place in the world and the things that can and should be done. Moreover, the desire to work out details of the scientific world-picture has an important motivating value for doing science. For the realist, the fact that a scientific world-view can have such force in practice is one more argument that realism is part and parcel of science.

### 3.3 Physical monism

A realist interpretation of science does not automatically involve the monistic idea that the world consists of only one kind of substance. But in actual fact realism and monism very often go together. This is because for the realist relations between theories are relations between different layers of reality; and from that perspective modern science itself provides grounds for monism.

To make the line of argument clear, consider the following. Physics can treat phenomena which also fall within the province of chemistry:

calculations in terms of simple chemical molecules can be replaced by calculations in terms of physical atoms. From a realist point of view this not only means that the observable phenomena can be attacked with both theoretical schemes, but also that the (really existing) molecules actually consist of atoms and that it is possible to calculate their properties from the properties of the constituent atoms. The conclusion naturally follows that physics pertains to a more fundamental level of reality than chemistry.

Admittedly, it is not possible to reconstruct the whole of chemistry from physics. But some incursions into the territory of chemistry *are* possible, and more is becoming possible with growing calculational power; there is no indication that there is a limit of principle to the validity of physics within chemistry. Very similar things can be said about the relation between biology and chemistry. An important area within biology, molecular biology, can actually with equal justification be considered as a part of chemistry. In other words, there are well-confirmed empirical points of contact between physics and a number of other disciplines; and also between physical subdisciplines. Within the realist interpretation of science this very cogently points in the direction of monism: ultimately, the whole world consists of elementary particles, and for this reason elementary particle physics deserves the title of the most fundamental description of the world.

Recent developments in elementary particle physics itself add another dimension to the picture. There are hopes that a unified theoretical basis can be achieved for all basic physical processes; a basis consisting of a very small number of principles. According to the realist, the explanatory efficiency of such a Theory of Everything would justify the belief that the theory describes the most fundamental level of reality.

Adherents of the idea of an ultimate physical Theory of Everything usually hasten to add that in practice it will be impossible to give really detailed explanations, on the basis of the theory, for phenomena outside elementary particle physics. But they stress that nevertheless from a philosophical point of view the theory makes it clear that everything originates in the processes going on at the most basic physical level.

## 4 Present-day anti-realism

What do modern sceptics set against such realist views? Of course, the most immediate reaction is to deny that science is an enterprise whose aim is to give an objective description of an independent world. We have already seen the traditional empiricist motivations for such a position. In recent discussions these have often been combined with a second line of argument. Is the realist idea that our theories conform more and more to an independent reality not an absurd underestimation of the role played by our own ways of looking and interpreting?

The theme of this question goes back at least to Kant, who held that we can never have access to things in themselves (the 'noumena'), but only to things as perceived, filtered and interpreted by us (the Kantian 'phenomena'). He concluded that it is a mistake to think that empirical science studies a world in itself; only 'phenomena' (in the Kantian sense) are the object of science.

Somewhat similar notions emerge in the work of Kuhn (and Feyerabend) from the 1960s. According to them, science is shaped by conceptual schemes, 'paradigms', ways of looking at the world, which can be stable for a long time within a community, but can also suddenly change. The success of a scientific theory therefore reflects not simply its adequate representation of the world, but rather the way scientists approach the world.

A currently popular way of arguing for the same conclusion is 'social constructivism'. Social constructivists contend that scientific theories, and scientific world-pictures, first of all mirror the social relations among the scientists who construct the theories and pictures. Scientific knowledge is regarded as the outcome of a process of negotiation between scientists, in which questions of prestige and power are paramount. In the most far-reaching version of social constructivism it is claimed that science is completely determined by such social factors, and cannot at all be said to reflect the characteristics of an independent physical world. Elementary particles, for example, cannot then be considered to exist independently of the scientist: they are literally *constructed* by the researchers themselves. The idea is that after a complicated process of negotiation, consensus is reached in the scientific community about the outcomes of certain experiments and their interpretation – for instance that the results will be spoken of as indicating the existence of quarks. But this final situation only reflects the power relations and interests in the scientific community.

It seems that in this extreme form the position cannot easily be maintained. Even if social struggle plays the important role that has been suggested, this does not imply that the outcomes of experiments can always be plausibly interpreted in a socially desirable way. This is already true for the outcome of a single experiment; but it is even clearer for a series of repeated experiments, in which parameters can be varied and disturbing influences can be controlled. It seems evident that in this case the facts should *allow* the proposed interpretation. Furthermore, if the empirical facts allow some socially desirable interpretation, so we could say that the *interpretation* is a social convention, it is not easy to see how it can also be a convention that the same phenomena repeat themselves in the same conditions. The standard reply of the social constructivist is that the decision of what are and what are not the 'same phenomena' and the 'same conditions' is *itself* the outcome of social negotiations, and not something which is immediately given to us. There is certainly some truth in this (see below). But it remains the case that once such negotiations have been concluded, it is possible to repeat experiments in a completely standardized way, without human intervention, with each time the same type of results. It seems therefore inevitable to assume that there is *something* outside us, with its own independent characteristics, something which *allows* the interpretations that we actually give.

Less extreme versions of the social constructivist idea accordingly appear more plausible. Here it is admitted that scientists *do* engage Nature, but that a complicated process of 'learning' to see order in the chaos of phenomena is part of experimental practice.[1] This process of learning is not the passive recording of data; the scientist plays an active part and to some extent indeed constructs the facts himself. Scientific observation is not just looking at phenomena, but involves the 'construal' of meaningful facts. The scientist has to classify phenomena, develop and apply concepts, decide about what is essential and what is 'noise'. In this process of construal social factors will be important, and cultural resources (including prevalent world-views) will be drawn upon to provide the pictures and terms to deal with Nature. Scientists thus construct facts; but Nature sets limits to the possibilities of construction. Not every construal, no matter how socially desirable, can be implemented.

Even a rather moderate and plausible point of view as just sketched poses problems for the realist position that science provides objective and universal world-pictures. These pictures are now taken to reflect the perspective, the intentions and the background of their founders as well as Nature. If we had been brought up in another culture, with a different world-view, different needs and interests, our scientific view of the world could well have been very different. Science is considered as an *interpretation* of Nature, which is dependent on, and in turn also influences, broader perspectives on the world.

### 4.1 Modern pluralism

More or less arguing along the lines of the foregoing, and adding that the superiority of science over non-scientific activities should not be assumed *a priori*, Paul Feyerabend has for a long time been defending the position that there should be no *privileged* status for a scientific world-view.[2] Non-scientific views construe the world in their own terms, and should be judged according to their own criteria. Moreover, Feyerabend claims, if we take a close look at science itself we see that it is actually a mistake to think that there is one all-embracing scientific world-picture. Different disciplines possess their own methods, theories and favourite pictures. Even within physics, there is a vast difference between fields like high-energy physics and – let us say – non-equilibrium thermodynamics. There is hardly any point of contact between these fields. Why would one want to say that one is more fundamental than the others? Each one serves its own purpose, just like countless other human activities serve their own purpose. Our lives consist of a patchwork of such different activities, with different motives and perspectives. None of them is 'universal', and it cannot be said that any one of them is 'fundamental' and founds the others.

In practice, we have to admit the existence of a plurality of methods, theories and pictures within science. However, one can argue that this by itself does not yet show that all such approaches have the same status. Confining ourselves to physics, we have already encountered the idea that elementary particle physics is at least in some sense more fundamental and universal than other branches of physics. This is not in the sense that other fields could actually be replaced by elementary particle physics, or that elementary particle physics gives more useful results. If practical usefulness is at stake, thermodynamics has undoubtedly a higher status. The intended sense in which elementary particle physics is taken to distinguish itself, is that everything in the world can be taken to consist of elementary particles, and that all phenomena are the eventual outcome of the interactions between them. As stressed in Section 3, this is not a gratuitous assumption, but is backed up by the existence of relations between particle physics and other fields. How can the modern pluralist defend himself against the reductionist attack that comes from science itself?

One possible response is to accept the realist outlook and to admit that physics is fundamental and universal in a very abstract ontological sense, while still denying that this means that other disciplines could be reduced to physics. This position is well-known from discussions about the relative autonomy, with respect to physics, of disciplines like biology and psychology. In biology one can reject the notion that non-physical 'vital forces' operate in living organisms, and at the same time oppose the idea that biology is 'nothing but physics'. The key idea is that biological terms and explanations derive their meaning from the conceptual schemes which are used *within* biology; they often have no meaning in physics. If this point is given full weight, it is possible to be an anti-reductionist, even if one concedes that the physical description can be interpreted using a realist approach, is complete in its own terms and moreover that biological and psychological characteristics 'supervene' on the physical level (which means, loosely speaking, that the physical state description completely fixes the state as described in biological or psychological terms). The existence of such a supervenience relation does not entail that biological or psychological terms correspond to something with an independent physical meaning (that is, to physical concepts which can be defined without recourse to the supervening biological or psychological concepts). And even if there were such relations between concepts of the different disciplines, the biological and psychological descriptions serve purposes which are totally different from the purposes of the physical description. There is no sense in replacing a psychological treatment by one in terms of elementary particles.

Such a response is monistic in its basic ontology, but pluralistic in all other respects. The various scientific disciplines are considered as looking at the world from different angles; the resulting pictures complement each other, and are not in mutual conflict. Reduction to physics only makes sense *insofar as one chooses to adopt a physical standpoint*; for example if one regards an organism as a physical system. By definition this does not lead to a reduction of biology *as such* to physics or chemistry. From this standpoint a physical Theory of Everything is a contradiction in terms.

### 4.2 Scientific agnosticism

In the position just sketched the fundamental ontological significance of physics was accepted. But a pluralist like Feyerabend will not be inclined to give in that much to the realist. He can draw support from the work on the social context of science which was mentioned before, and in which it is made plausible that scientific pictures reflect not only Nature but also ourselves. He can also take comfort from the fact that there is a modern form of empiricism, developed by Bas van Fraassen,[3] which is able to answer the objections to earlier empiricist philosophies of science (especially logical empiricism).

As for earlier empiricists, the aim of science for van Fraassen is to save observable phenomena, not to discover Truth about unobservable things. But he concedes to the realists that the distinction between observation terms and theoretical terms (as made by the logical empiricists) is not tenable. In van Fraassen's view *all* terms in scientific theories *could* refer to really existing entities and processes, regardless of whether the things referred to are observable or not. All terms have the same linguistic status; there is no distinction between a list of observation terms which really refer and a list of theoretical terms which only obtain their meaning via relations with the observation terms. But the important empiricist point remains, says van Fraassen, that some parts of the world are observable (because they are big enough to be seen, or have an interaction with us that is strong enough), whereas others are not. (Clearly, this is not a linguistic distinction, but a physical one.) Therefore, although the world *could* be the way our theories say it is, we do not have sufficient evidence to assert that it actually is. We cannot *know* whether there indeed is a correspondence between the terms of our theories and the unobservable parts of the world, because we are unable to check that correspondence empirically. The point is strengthened by the fact that there can be many rival theories that save the same observable phenomena; how could we ever decide which one of them is true or comes closest to the truth? It seems that we must remain agnostic about such things.

Van Fraassen criticizes the realist proposal that 'inference to the best explanation' can provide us with a clue about which theories best describe the world as it really is. He points out that the explanations which we actually possess and appraise are certainly only a small subset of all possible explanations. The chance that the *true* explanation is among the ones we are actually contemplating must for that reason alone be very small. Further, it is hard to see how some of the features that make for a better explanation (greater content of information, for example) can make it more probable that the explanation is true. The more information a statement contains, the more specific it is, the greater the chance that it is false! We may add that it can also be argued (see the beginning of Section 4) that there are significant social influences on what we value as good explanations. This sheds doubts on the idea that what makes an explanation good or bad for us is determined by its relation to the world (its 'truth-content').

According to van Fraassen the best response is to be a philosophical agnostic; we are simply not in a position to know how the world is. The aim of science is merely to discover empirically adequate theories. Acceptance of a theory as empirically adequate does not imply belief in what the theory says about unobservable parts of the world. For an empiricist like van Fraassen the endorsement of a theory is therefore compatible with agnosticism about unobservable things, even if the theory in question explicitly makes statements about these things. This is not to deny the fact that world-views derived from science play an important role in practice. But according to van Fraassen the right scientific attitude is to criticize such world-pictures, and to point out that alternatives are possible, rather than to subscribe to any particular 'scientific world-view'. This position is reminiscent of Mach's insistence that the essence of science is in its critical methodology.

What consequences does an agnostic view have for the issues of whether physics is fundamental and whether there is room for other approaches alongside physics? Of course, relations between physical descriptions and other descriptions as they are empirically found to exist cannot be denied on any interpretation of science. But in an agnostic view they can be taken to be exactly what they are found to be: relations between observable consequences of theories. Their existence implies that there are correlations between the predictions of different theories. But such connections need not mean that one theory describes a level of reality which is more fundamental than the others – we can remain agnostic about *any* reality behind the phenomena. According to the sceptical standpoint theories do not aim at fundamental descriptions of reality at all.

Therefore, the existence of a fundamental physical world-picture is something very natural only in the realist view of science; there its possibility is built in from the beginning. In an agnostic view (which can be combined with the notion that different disciplines possess their own autonomous perspectives and 'complement' each other in the way sketched in Section 4.1) the idea of a physical Theory of Everything becomes very far-fetched.

## Notes

1. David Gooding, *Experiment and the Making of Meaning* (Kluwer, Dordrecht, 1990). This also gives other references to the work of social constructionists.
2. Paul Feyerabend, *Against Method*, (New Left Books, London, 1975); *Farewell to Reason* (Verso, London, 1988); *Three Dialogues on Knowledge* (Blackwell, Oxford, 1991).
3. Bas van Fraassen, *The Scientifc Image* (Oxford University Press, 1980); *Laws and Symmetries* (Oxford University Press, 1989).

## Suggested further reading

1. Ian Hacking, *Representing and Intervening* (Cambridge University Press, 1983).
2. Barry Barnes, *Scientific Knowledge and Sociological Theory* (Routledge and Kegan Paul, London, 1974).
3. Jarrett Leplin (ed.), *Scientific Realism*, (University of California Press, Berkeley, 1984).

# 4 Enlarging the known world

Ernan McMullin

It is a commonplace that the natural sciences 'enlarge' our world, that they enable us to reach outside the narrow circle of our sense-knowledge to more complex domains of many layers. But it is not so simple to specify in what this enlargement consists. Most would say that we come to know of the existence of myriads of entities of which our ancestors knew nothing. But how exactly do we *do* that? And how reliable can such knowledge-claims be? Questions like these continue to divide philosophers of science. Conventionally, the main division is said to lie between 'realists' and their critics ('anti-realists'). But there are, notoriously, almost as many realisms as realists. And the critics of realism represent a wide variety of philosophical positions. So boundary-lines shift, and differences that at first sight appeared fundamental vanish as the debate continues. Still, there is in the end a genuine disagreement here, and it concerns the most important philosophical question that can be raised about the significance of the natural sciences: what quality of understanding do they afford of the underlying structures of the world around us?

After a preliminary discussion of the tangle of differences separating realism and anti-realism, I shall argue for a broadly realist answer to this last question, taking care to avoid the overstated versions of realism on which critics have too often focussed, versions that they have in some instances themselves created. I shall assume that the disagreement is at bottom one about *existence*. Realism bears on what can be claimed as real, that is, on what can reliably be claimed to exist. Issues regarding truth and reference are relevant, but are not the focus of the debate.

Directly opposed to realism, therefore, is instrumentalism, the view that the theoretical constructs of the scientists function only as instruments of prediction; they do not serve to define and discover things of whose existence we were previously unaware. The aim of theory is predictive accuracy, not understanding. It does not enlarge the world, in the sense of bringing new levels into view; it leaves the furniture of the world as it was but affords new control over it. The disagreement between the instrumentalist and the realist is unequivocal. Realists of all varieties affirm that theory serves to enlarge the world-as-known, to discover new and unsuspected variety and complexity within it; instrumentalists tend to regard 'the appearances alone as real',[1] and all else as construction, as a means to guide us on our path, no more.

But is there a real debate? Are there, in fact, any instrumentalists? The position seems such a counter-intuitive one that one's immediate inclination is to suspect that the realist may be conjuring up bogeymen. True, there have in the past been notable defenders of instrumentalism in the context of mechanics; Berkeley and Mach come immediately to mind. And one may once again today find instrumentalist views among those who regard quantum mechanics as the paradigm of science. But when it comes to such mainline areas of science as cell biology, astrophysics, or geology, where tentative existence-claims based on explanatory success are routine, few philosophers of science seem eager to defend the stark instrumentalist line.[2]

What is odd, then, about the present situation is that the most energetic critics of scientific realism go to some lengths to distance themselves from the instrumentalist alternative. They are united in disapproval of the arguments used in support of scientific realism, while being wary of being associated with the contrary view. But theirs is not a simple withholding of assent. Many of them maintain the viability of an intermediate position, like Arthur Fine's 'natural ontological attitude', which allows them to admit the existence of some (or even all) of the entities that scientific realists propose while deprecating the manner in which they propose them. (An unsympathetic outsider might be disposed to comment that they are enjoying the advantages of theft over honest toil!) If their answer to the familiar test-question: 'Do electrons exist?' is: 'Of course, they do!', they are allied with the realist cause, whether they like it or not, and as dependent as realists are on finding a warrant for the confidence that most scientists would manifest in that answer.

## 1 Ampliative inference

One way – admittedly not the usual one – to describe the debate regarding scientific realism is to say that it concerns the nature and limits of ampliative inference as this is found in science. In deductive inference, rules are applied to extract from the premisses what is already implicit there. There is no element of risk involved; the conclusion does not extend the slightest fraction beyond the domain from which one began. In ampliative inference, by contrast, a risk is taken, a venture is made beyond the strict limits of the original evidence. In the natural sciences, the form the risk usually takes is that one goes beyond the data of observation to lay tentative claim on what will be found in a domain that in one way or another transcends these data. In the strictest of empiricist regimes this would not be permitted, but such a regime would have to outlaw even the simplest sorts of empirical science.

Ampliative inference takes two main forms in natural science. The first, inductive generalization, is only weakly ampliative. From a few instances one infers tentatively what will be true of a class; from a limited regularity one reasons to a larger one; from a handful of data-points one constructs a curve; from a sample, one reaches out to make a statistical assertion about an entire population. In all these cases, what is observed is taken to be representative of a larger class, itself not observed. There are obvious risks involved in this sort of extrapolation: that the sample may prove unrepresentative, for instance, or that the generalization arrived at might be 'accidental', that is, might not reveal the sort of lawlikeness on which further inference could reliably be based.

But the epistemic risks entailed in the second kind of ampliative inference are much greater. Instead of terminating in an empirical regularity, universal or statistical in form, as induction does, retroductive inference takes such regularities as a starting point and purports to explain them by postulating underlying processes, structures, entities, causally sufficient to account for them. Thus instead of calling for more of the same as induction does, retroduction typically introduces new sorts of causal agency, new in the sense that they differ in kind from the effects to be explained. This kind of inference requires in the first place a degree of imagination; it is not a matter of rule or generalization. The warrant for it lies in the quality of the explanation given.[3]

C. S. Peirce's distinction between these two forms of ampliative inference is familiar. The two are, as he reminded us, intertwined in actual scientific research. Induction provides the material for retroduction to work on; the effects that scientists seek to explain are either explicitly lawlike in form or are at least reproducible, i.e. implicitly lawlike. (Historical sciences, like paleontology and cosmology, raise special issues.) Despite what much of the recent philosophical discussion of explanation in science appears to assume, explanation in the natural sciences is not ordinarily of singular events. On the other hand, induction may also require retroduction as underpinning; if there is a question as to whether an inductive generalization represents a genuine law or not, the standard test is to inquire into the plausibility of its yielding a theoretical explanation.

Instrumentalists are likely to blur this distinction between induction and retroduction; characteristically, they will tend to lump all forms of ampliative inference together under a simple label, 'induction'. For them, retroduction does not yield an inference to unobserved entities; it is not a causal explanation, in the strong non-Humean sense of 'cause'. It is simply a more complex form of induction, involving theoretical construction, it is true, but only as a means to more effective prediction. It is explanatory only in the weak Humean sense which permitted the deductive–nomological (DN) model of explanation to imply that explanation and prediction are two sides of the same coin.

## 2 Realism and anti-realism

The disagreement between realism and anti-realism is primarily about retroduction, because retroduction is the primary means whereby the natural sciences enlarge our world, according to the realist viewpoint. The world of the instrumentalist is a shrunken one, precisely because retroduction is *not* accepted as ampliative. The issue is not in the first place about induction, about whether or not there are 'necessities' in Nature, about whether there are genuine 'laws of Nature', important though these issues are. These are second-order, partly metaphysical, questions about the proper interpretation to be given to the regularities on which the natural sciences are grounded; the realist will typically defend a strong version of lawlikeness, while the instrumentalist will tend to deny that there are anything like laws of Nature or natural necessities.[4] But this difference must not be allowed to distract from the main issue separating realists from their critics: does retroduction serve to enlarge our world or not?[5] Can it afford a tentative warrant for affirming the existence of the theoretical entities it introduces to explain observed regularities?

It was in the 1960s that Kuhn and Feyerabend began to develop the theme of the instability of theory in science: the history of science seems to show that theories come and go, sometimes in abrupt and discontinuous ways. Hence, one cannot take seriously any existence claim, however weak, for the entities and processes postulated by theory. This is the argument that recent critics of scientific realism have most relied on. The instability argument is directed against a realist understanding of retroduction in science generally, more specifically against what Gilbert Harman called 'inference to best explanation'.[6] Hence, it applies as directly to theory in geology or astronomy as it does to theory in quantum mechanics. If this is the argument the critic is relying on, there should be no distinction between theoretical entities, so to speak; the critic who rejects an existence claim on the grounds of theory success for electrons should do the same for galaxies. If the potential for theory instability undermines one, it does the other. If one wishes to maintain that there are good grounds for supposing the existence of galaxies as vast collections of discrete stars, one ought to be willing to allow existence claims on the same sorts of grounds elsewhere.

What has further muddied the waters in the last decade or so has been the growing critique of a variety of metaphysical views said to be associated with realism. Hilary Putnam, a strong defender of scientific realism in the early 1970s, went on to distinguish between a 'metaphysical' version of realism which he rejects, and an 'internal' version which he is still willing to accept. Metaphysical realism he defines in terms of three theses: (1) the world consists of a fixed totality of mind-independent objects; (2) there is exactly one true and complete description of the way the world is; (3) truth involves some sort of correspondence.[7] While it is true that many defenders of scientific realism might be disposed to defend one or other of these theses, particularly the third, nothing compels them to do so. Though the same term 'realism' was still used, Putnam was really changing the topic of discussion. Scientific realism bears on the existence claims that are made for the sort of entities that are typically postulated in scientific theory. Putnam's concern was no longer with this issue but with a much broader set of questions bearing generally on the relation between language and world, particularly on the nature of reference. His strictures on metaphysical realism bear on the term 'cow' (the term he often uses by way of example) just as much as they do on the term 'electron'.[8]

Someone who doubts whether the world consists of a 'fixed totality' (Putnam's phrase) of mind-independent objects would not thereby be prevented from holding that the explanatory success of a scientific theory gives us a (qualified) reason to believe that the entities postulated by the theory actually exist. Putnam's polemic against the 'God's Eye View', the 'One True Theory', is prompted by a Kantian conviction that it makes no sense 'to ask whether our concepts 'match' something totally uncontaminated by conceptualization'.[9] It is a property of the world, he says, that it admits of different mappings; there is no theory-independent 'way the world is'. The list of things that exist must always remain theory-relative, the term 'theory' to be understood in the broadest possible sense. (Cows are, for example, part of our everyday 'theory' of the living world.)

But it is none the less objective for that, Putnam assures us. The function of his 'internal realism' is, in fact, to secure the objectivity of his proposal in the face of the charge of relativism that his refusal to allow correspondence of any sort into his notion of truth might seem to warrant.[10] Internal realism is a theory of truth, not a form of reductionism, he insists; it is not a dispute over what really is, but about how reference functions.[11] Although internal realists maintain that nothing we say about an object describes it as it is in itself independently of its effect on us, they do not doubt that the objects their terms refer to ('things-for-us') do in fact exist, though admittedly not in the (proscribed) sense of 'exist' favored by the metaphysical realist.[12] Putnam's internal realism does not, then, seem to inhibit the enlargement of the world that the scientific realist affirms, though his idealistic perspective makes his 'world' a more shadowy place than the one we have been accustomed to.

This is not, of course, to say that internal realism includes scientific realism as a special case, or anything of the sort. Indeed, some of the arguments Putnam uses against the correspondence theory might be thought to imply the very opposite. A keystone of these arguments is his claim that 'you cannot single out a correspondence between our concepts and the supposed noumenal objects', an access he takes it as obvious that we cannot have.[13] And in support of this claim, he cites the example of Newtonian physics:

> If Newtonian physics were true, then every single physical event could be described in two ways: in terms of particles acting at a distance, across empty space (which is how Newton described gravitation as acting), or in terms of particles acting on fields . . . which finally act 'locally' on other particles. . . . The Maxwell field theory and the retarded potential theory are incompatible from a metaphysical point of view, since either there are or there aren't causal agencies (the 'fields') which mediate the action of separated particles on each other (a realist would say) . . . If all it takes to make a theory true is abstract correspondence (never mind which), then incompatible theories can be true. To an internalist this is not objectionable: why should there not sometimes be equally coherent but incompatible conceptual schemes which fit our experiential beliefs equally well? If truth is not (unique) correspondence then the possibility of a certain pluralism is opened up.[14]

Here the attack on metaphysical realism seems to carry with it a rejection of scientific realism also. To say that it is only from a 'metaphysical' point of view that one can claim that there either are or are not causal agencies mediating the action of separated bodies on one another would seem to render 'metaphysical' the kind of affirmation of existence that scientific realists customarily make, thus denying them legitimacy from Putnam's standpoint. He asks, why should there not sometimes be equally coherent but incompatible conceptual schemes which fit the data equally well? Why not, indeed? An even better example might be drawn from contemporary quantum mechanics where the incompatibility of the two leading interpretations, the indeterminist one of Bohr and the determinist one of Bohm, has recently been receiving a great deal of attention.[15]

Situations like these do not, it should be said, undermine the credibility of the realist construal of theory in general. For one thing, the inability at a given moment to decide between two such alternatives is not necessarily permanent, in that the two interpretations may prompt rather different responses to new data, leading to a modification of the original formalism. This happened in the Newtonian case, in fact, with the introduction of the field concept. More importantly, what we see in cases such as these are two different interpretations of the *same* formalism, not at all the normal situation where two *different* theories and their accompanying formalisms account equally well for the available data. The moral for the scientific realist in the former cases (which seem to occur almost exclusively in mechanics) is that the matter of *interpreting* a formalism in order to decide what the ontological implications of a particular theory are, may sometimes be far from straightforward; indeed, it may be indeterminable.

What is especially significant about the position of the 'internalist', as Putnam outlines it above, is the sort of pluralism that would apparently allow one both to hold and to deny that causal agencies are at work mediating actions of separated bodies on one another. But perhaps Putnam is only reminding us of the sometimes paradoxical consequences of adopting a non-correspondence notion of truth; perhaps he would still allow the truth of qualified existence-claims for some theoretical entities, provided that 'truth' be interpreted in his way. Or would he exclude such claims on the grounds that they assert privilege for one particular way of cutting up the world? It is hard to be sure. What *is* sure is that the focus of his attack on 'realism' is not scientific realism, as that position is defined here. His attack might, however, have negative consequences for scientific realism also, depending on how far his Kantian sympathies extend.

### 3 Realism and pragmatism

Something similar might be said about Richard Rorty's critique of realism. The term 'realism', he remarks, 'has come to be synonymous with anti-pragmatism', that is, with opposition to the approach to philosophy he advocates, dubbed 'pragmatism' by him because of its resonances with the philosophy of Dewey and James.[16] (It might be more apt, perhaps, to say that pragmatism for many has come to be synonymous with anti-realism.) The generic 'realism' he rejects he associates with the correspondence notion of truth, a notion that in his view has defeated all attempts at elucidation, including Kripke's efforts to construct a physicalistic theory of reference. He links realism more generally with Philosophy (capitalized), the tradition of inquiry into truth and goodness he traces back to Plato, a tradition now in his view discredited. Within Philosophy there has been, to be sure, a deep division about the nature of truth between those who see truth as transcending the sensible world of space and time, and those who insist that since the space–time universe is all there is, truth amounts to a correspondence with material reality. For the latter (the 'positivists'), natural science is all the basic truth there is; for the former (the 'Platonists'), science takes second place to metaphysics.

Pragmatists set both aside as 'Philosophers' who assume that 'the assemblage of true statements should be thought of as divided into an upper and lower division, the division between (in Plato's terms) mere opinion and genuine knowledge'.[17] For the pragmatist, on the other hand, true propositions are not true because they correspond to reality: 'modern science does not enable us to cope because it corresponds, it just plain enables us to cope'.[18] This *sounds* like instrumentalism, to be sure, but Rorty lumps instrumentalists with scientific realists, dismissing instrumentalism as 'a quaint form of late Platonism'.[19] Both sides are committed (he claims) to the view that there is an abductive (in my terms, retroductive) method which is distinctive of natural science, making science a sort of natural kind, possessing its own essence.[20] Scientific realists base their argument on the supposed virtues of this form of inference; instrumentalists rest their position on a critique of it. Both sides also (according to Rorty) require a peculiar 'Philosophical' sense of 'existence' in order to make their point.

I would question these characterizations. Scientific realists do not have to suppose, so far as I can see, that the form of inference that is at the heart of science (retroduction) is peculiar to science. True, it is more developed, more circumscribed, there.[21] But when someone infers the existence of a mouse in the wainscoting as the explanation of small sounds in the night (to turn an example of van Fraassen's around), the same *general* sort of inference is being employed that a physicist might use in interpreting cloud-chamber data. Furthermore, when a physicist identifies a particular ionized track as that of an electron, the existence claim implicit in 'an electron caused that track' is of just the same kind as that implicit in 'a mouse made that noise'. When a scientific realist (in the face of empiricist objections) infers the existence of electrons in the light of all the evidence available, the term 'existence' is not being used in some special 'metaphysical' sense, but in the same sense in which someone might say: '*Of course* there are mice'.[22]

Rorty's fire is directed against 'Philosophers' who attempt to *explain* the success of scientists' theories by postulating the existence of the appropriate sort of entities:

> From a Wittgensteinian or Davidsonian or Deweyan angle, there is no such thing as 'the best explanation' of anything; there is just the explanation which best suits the purpose of some given explainer. Explanation is, as Davidson says, always under a description, and alternative descriptions of the same causal process are useful for different purposes. There is no description which is somehow 'closer' to the causal transactions being explained than the other.[23]

On the contrary, there often *is* a description which can claim to be closer to the causal transactions than others, namely, the theory generally adjudged best in the light of the evidence available in that part of natural science where transactions of this sort are discussed. Making that judgement is precisely what theory evaluation in science is all about. Invoking the pragmatic aspect of explanation, underlining the fact that explanation is of its nature context-relative, does nothing to lessen the force of this rather banal reminder. The consolidation over time of a particular scientific specialization like organic chemistry or population biology creates what might be called a *standard* context of evaluation, standard, that is, relative to the current practice of that specialization. Rorty's phrase, 'what best suits the purpose of some given explainer', gives a misleadingly particularist cast to the evaluation of explanations (theories) in science.[24]

That is not to say, of course, that organic chemists are likely to agree, routinely, on what is to count as the best causal account of the data at hand. Kuhn has pointed to the plurality of epistemic values involved in theory appraisal (of which more below); different understandings of these values on the part of individual scientists, different relative weights attributed to them, may lead to disagreement, sometimes long-lasting disagreement.[25] But the particularity of the individual theory evaluation is modulated in complex ways by the social, as well as the temporal, character of scientific inquiry. C. S. Peirce, who, incidentally, counted himself a pragmatist, long ago pointed out just how this comes about.[26]

Most philosophers of science today would probably agree that theory evaluation is not a straightforward matter of applying logical rules, that a formal confirmation theory of the sort the logical positivists sought is a chimera, but they would also hold that this does not compromise the long-term objectivity of scientific inquiry itself.[27] Admitting the perspectival character of theory evaluation by the individual scientist does not lead to the kind of multiplicity of 'alternative descriptions of the same causal process . . . useful for different purposes' (in Rorty's words) that proponents of pragmatic theories of scientific explanation like to conjure up. Organic chemists who are trying to explain a particular process do not work at cross-purposes. They can appeal to more or less common standards, standards to which referees of their work can also appeal. Although this does not guarantee unanimity, it does ensure that 'alternative descriptions of the same causal process' do not simply remain as incommensurate alternatives but compete with one another.

'What is so special about prediction and control?' Rorty asks. 'Why should we think that explanations offered for this purpose are the "best" explanations? . . . What is the relation between facilitating prediction and control and being "non-perspectival" or "mind-independent"?'[28] Understanding Nature presupposes an accurate knowledge of the regularities which lead us to speak of 'Nature' in the first place. Accurate description is constitutive of the activity we call 'natural science', and accurate prediction is thus a test, indeed the primary test, of what counts as the best (current) causal explanation of a natural regularity.[29] A theory that fails this test is deficient as causal explanation. Requiring this test constrains the perspectival character of evaluation; differences in perspective have to be overcome (not always an easy task) in order that epistemic values, such as predictive accuracy, coherence and fertility, may as far as possible be maximized. Thus, the pluralism evoked by the pragmatist does not, in the end, seem to threaten the realist's reliance on retroduction, on the processes leading to claims of 'best' explanation.

Whether someone who rejects the claims of scientific theory to constitute (within the limits sketched above) the best available causal explanation can avoid the label of instrumentalism seems to me doubtful. It may well be that instrumentalists in the past have accepted the correspondence view of truth and other 'Philosophical' presuppositions rejected by Rortyan pragmatists. But on the crucial issue of whether the explanatory success of a theory supports to some degree belief in the existence of the sort of entities the theory postulates as causes, the pragmatist appears to stand with the instrumentalist. I noted earlier that critics of scientific realism generally attempt to distance themselves nowadays from instrumentalism (or even, as in Arthur Fine's case, from 'anti-realism'). But if scientific realism be detached from the various metaphysical theses discussed above, it seems to me that these claims to have discovered a third 'neutral' standpoint beyond realism and instrumentalism must fail. There does not seem to be a viable third standpoint: one either accepts the claim that retroductive success has legitimate existential implications or one does not.

## 4 Realism and empiricism: historical notes

On the face of it, there would seem to be no reason why someone who claims that a strongly-supported retroduction can warrant the existence of the entities it postulates should not also be an empiricist. Yet van Fraassen, seeking a more positive label for his view than 'anti-realism', names it 'empiricism'. His constricted world he describes as the world of the empiricist. How is this sort of 'empiricism' to be understood? Must defenders of scientific realism renounce empiricism if they are not to risk inconsistency? It may be worth making an extended historical excursion first before continuing with our exploration of the world of the realist.

Aristotle's emphasis on the primacy of experience (*empeiria*) is well-known. Whatever is in the intellect, he insists, must have come to it through the senses. There are no innate ideas, no recollections from an earlier existence; the intellect cannot arrive at first principles out of its own resources. In this sense, then, Aristotle is an empiricist, indeed a paradigm of empiricism. Yet he is equally a realist, quite assured that we have a secure knowledge of the world as it is in itself. According to him, what enables us to attain this is, first, that the forms which make things be of the kind they are are conveyed to the mind through the senses; the form in the mind and in the thing are in a real sense the same. Second, through a sort of skilled insight (*epagōgē*) we can come to grasp necessary connections between essence and property or between property and property. Thus science can be said to be 'of the necessary'.

What is meant by 'necessary' here? Did Aristotle believe that everything behaves inexorably as it does? Clearly not. Human action is not, in general, necessitated; the ability freely to choose a course of action is central to Aristotle's ethics. More important for our present topic, 'chance' events are (according to Aristotle) of frequent occurrence in the world around us. When forms are realized in matter, the individual so composed is subject to interactions with other similarly matter-bound individuals. The outcomes of these interactions may thus, in effect, be unpredictable. The teleological reasoning that assures us, for example, that acorns tend to grow into oaks, does not exclude the possibility that in the messy circumstances of material existence in place and time, some acorns might shrivel by the wayside and others might be eaten by pigs. There is no failure here in the necessity by which acorns 'naturally' become oaks and not maples; this necessity is hedged by all sorts of *ceteris paribus* conditions, that is all. No indeterminism need be involved in chance events of this sort; it is because there is an intersection of multiple *independent* causal lines that the outcome is said to be 'chance', not because there is an indeterminacy in any one of the lines.[30] How can an empiricist *know* all this? That a certain kind of thing will continue to act as it has, for example, if not actively prevented? The so-called 'problem of induction' simply does not arise for Aristotle because the universal in his view can be grasped in the experienced singular.

The reaction that set in against Aristotelian natural philosophy among Christian theologians in late thirteenth-century Western Europe was prompted in part by this doctrine of 'necessities'. The universe was not to be viewed as a 'given', as it had been by Aristotle. Rather, it ought to be regarded as the work of an omnipotent Creator, one who is free to choose the sort of world that would come to be, and who is free also to interrupt normal process by way of miracle. Aquinas believed he could retain the Aristotelian doctrine of demonstration while introducing two extra constraints on the necessities this doctrine requires as basis: first, that they be contingent on God's choice of universe, and second, that they apply only to the normal process of Nature, ignoring miracles. These necessities themselves are still to be grasped as intrinsic to the forms conveyed through the senses; if this is correct, no ampliative inference is required, strictly speaking.[31]

Others were less sympathetic to Aristotelian doctrine. William of Ockham rejected its central pillar, the notion of a nature shared in common. There are only individual things and their sensible qualities. A term like 'horse' groups together things that resemble one another in certain respects; the individuals so designated are called 'horses' not because they share a common form or essence or nature, but because they are similar in specific respects. Such a similarity is still sufficient to support a science yielding certain knowledge of the physical world. It is based not on necessary relations between forms intuitively grasped but on regularities noted in our experiences of singulars. Induction, understood as generalization from sample to whole, is thus required. What allows an ampliative inference of this sort to conclude with certainty is (according to Ockham) an additional premise that similar individuals under similar circumstances will react in similar ways or exhibit a similar constellation of dispositional properties.[32] Supporting this premiss is a theological one that these uniformities are of the Creator's free and benevolent choosing; they are not due to the intrinsic constraints of Nature. Although they are contingent in the sense that God could have ordained otherwise and can at any time suspend the 'common course of Nature', they can nonetheless be confidently relied on in the guidance of action.

In van Fraassen's estimation, the nominalist philosophy of the fourteenth century 'marks the true birth of empiricism',[33] empiricism, that is, of the kind he identifies with. My purpose in making this excursion is to raise a question about this identification, and about the further implications of it for the issue of anti-realism. Ockham's empiricism is sometimes described as 'radical', in the sense that he not only (as Aristotle does) limits knowledge to what can be gained from the senses but also eliminates whole classes of entities, like common natures, on the grounds of economy. (Another label for this variety of empiricism might be 'eliminative'.) Ockham is suspicious of traditional metaphysics, suspicious of the tendency on the part of his opponents to associate entities with every abstract term. He reduces all qualities in physical things to qualities capable of being sensed by us, and stresses the primacy of quantity in the understanding of substances.[34] All of this would commend his brand of empiricism to van Fraassen.

There was, however, more to it. For a world of singulars to yield science, there had to be some sort of ordering principle. And there was. God's absolute power to bring to be anything not actually impossible is tempered by a freely chosen constraint, a benevolent commitment to a uniformity of behavior on the part of similar individuals, sufficient to warrant reliable inductive inference, indeed, even a limited form of physical necessity. The effect of this, of course, is exactly as though these individuals possessed the natures Aristotle had attributed to them. Aristotle's world and Ockham's world would *look* alike; the same configurations of sense-qualities, the same patterns of action, would be found in each. But the ontology of the two would be different, assuming that God's will that singular individuals behave in a certain way does not constitute an ontological *disposition* on the part of those individuals to act in that way.

What kind of empiricism is this? How radical *is* it? It is clearly eliminative in intent, but the resultant ontological deficit is made up by a definitely non-empiricist theological premiss. How bold, in the circumstances, is this premiss? Given Ockham's commitment to a strongly eliminative form of empiricism, a supplement seems to be needed if a proper science of the physical world is to be possible, as Ockham firmly believed it to be. The additional premiss could, therefore, be regarded as a reasonable hypothesis (certainly not the status that Ockham himself accorded it!) for a believer in the Christian God. Where else *could* one look for the needed supplement, after all? What saves Ockham's empiricism from its skeptical implications, therefore, is the theological framework in which it is set. When the eliminativist winds blew again in the early eighteenth century, this framework no longer seemed quite so obvious a support for the thoroughgoing empiricist.

The three major figures in the British empiricist tradition, Locke, Berkeley, and Hume, neatly exemplify the three broad possibilities in this regard. Locke was, in our terms, a scientific realist, unhesitating in his support for the prevailing 'corpuscular hypothesis', according to which the qualities of sensible bodies are caused by the minute particles of which these bodies are composed. Since the manner in which this subvisible action occurs can never be known by us with certainty, a 'science of bodies' is forever out of reach. He still equated 'science' with demonstration in the traditional way. But he does not leave it at that. A lesser form of judgement may still be accessible to us, if we are content (as we should be) in this context with probability.[35] Skillful employment of hypotheses based on analogies with the sensible world may still allow us access to the hidden realm of the corpuscles, though the judgements we make regarding this realm can be no better than probable. Locke makes it clear that as far as he is concerned this constitutes a genuine access, even though it goes far beyond the direct reach of the senses.[36]

He has a good deal to say about how hypotheses are to be assessed, much of it rather vague by our lights, but still exhibiting a shrewd grasp of where the difficulties lie. Hypotheses are to be judged in competition with one another on the basis of their own explanatory power (the 'greater light' they afford) and the explanatory power of the larger system of ideas of which they are a part. Initially, a hypothesis will always face some 'exceptions'; these are not to be taken too seriously until they are investigated in detail. After the available alternatives have been put to the test, one has to decide 'which is most consistent in all its parts, which least clogged with incoherences or absurdities, and which freest from begged principles and unintelligible notions. This is the fairest way to search after truth'.[37] What is significant here is Locke's realization that if one is to make claims about entities or events not directly accessible to our senses, the epistemic weight of enlarging the world in this way must be borne by the criteria one employs to evaluate the hypothesis involved.

Berkeley's response to the new science was quite different, in part because, unlike Locke, he took mechanics as his paradigm. The concept of force was the key to the new mechanics of Newton. But critics like Leibniz and the followers of Descartes were suspicious of Newton's attempts to *explain* motion by calling on forces, or more specifically, attractions. How, after all, does attraction operate? Action at a distance was unacceptable to all, and Newton himself had shown that there could not be a resisting medium bridging the gap between sun and planets. In the *Principia*, Newton had in consequence been studiously non-committal about the 'cause of gravity',[38] restricting himself to a 'mathematical' or descriptive account only, leaving aside the 'physical' issue of how planetary motion is to be *explained*.

Berkeley fastened on to this ambiguity. Notions like *force* and *attraction* 'are useful for reasoning and reckoning about motion . . . but not for understanding the simple nature of motion itself'. He takes Newton to have introduced attraction 'not as a true physical quality but only as a mathematical hypothesis', the product of 'abstraction' and clearly not 'an entity found in nature'.[39] 'Mathematical entities have no stable essence in the nature of things; they depend on the notion of the definer'.[40] This leads to a bolder claim: 'It is not in fact the business of physics or mechanics to establish efficient causes, but only the rules of impulsions or attractions and, in a word, the laws of motions'.[41] There is a weak sense of 'cause' which amounts to nothing more than regular succession, he allows; in this sense, and in this sense only, can physics be said to 'explain' motion.[42]

This is instrumentalism in one of its classical formulations. Yet Berkeley had no intention of leaving it at that. He was convinced that an impartial examination of Newton's mechanics would support an instrumentalist philosophy of natural science. But the importance of this lay elsewhere. A *proper* causal explanation of motion must be sought in a 'superior science' of spirit, that is, in metaphysics or theology. Since only spirit, in his view, can initiate motion, God must be the true cause of the motion of all inanimate bodies, from planets to falling stones. In this way, Berkeley avoided (just as Ockham had done) the skeptical consequences of his empiricist starting-point, but at the cost of a considerable departure from empiricism.

One other feature of his instrumentalism is worth noting. Though he takes it to apply to natural science generally, the case for it is in fact entirely dependent on his analysis of *mechanics*. It is the fuzziness he finds in the notion of force that allows him to dismiss force itself as an 'occult quality' or as a mere abstraction. The same argument would not work against Locke's corpuscles. If Berkeley's primary motivation was to present God as the ultimate source of motion in the world, it is not surprising, of course, that he would have focussed so exclusively on mechanics. He might have argued that *given* his theological premiss, retroductive inference to such hidden 'causes' as corpuscles would inevitably fail; no hidden physical causes seem to be needed – or indeed permitted – in such a system. But the argument carries much less conviction in this case than when he can argue from the plain deficiencies of mechanics as explanation to the *necessity* of the theological premiss.

It may be significant that in his late work, *Siris* (1744), Berkeley turns his hand to broader scientific issues, leaving the narrow track of mechanics behind. He inquires into the properties of bodies (beginning with 'tar-water', i.e. pine balsam, whose 'acid spirit' was held to have curative properties), and in the approved fashion of his realist predecessors postulates minute corpuscles, active spirits and an aether, as plausible explanations of the observed phenomena. Commentators on Berkeley have vigorously debated the consistency of all this with his earlier blend of idealism and instrumentalism.[43] My own perspective on the apparent shift is simple: once attention shifts from mechanics to the natural sciences in all their diversity, instrumentalism is likely to lose its appeal.

The last of our empiricist trio, David Hume, rejected both the realism of Locke and the theological premiss that protected Berkeley's instrumentalism (as it earlier had Ockham's nominalism) from its skeptical implications. From his starting-point in the theory of ideas he shared with his empiricist predecessors, he went on to defend a position that incorporated the elements of both nominalism and instrumentalism but went far beyond them to a skepticism so radical that it undercut his own attempts to give reasons in its support.[44]

His concern was not with the explanations offered by natural scientists of such things as planetary motion or barometric levels. These are for him pseudo-explanations since the presuppositions on which they rest – the reliability of induction and even the continuous existence of physical bodies – are themselves incapable of rational justification. His concern consequently is with the explanation of our *belief* in such irrational claims. Why *do* we, in the circumstances, continue to accept them? His answer is elaborate and ingenious (hence the interest in it on the part of contemporary philosophers) but ultimately, by most estimates, incoherent. Hume himself was well aware of the tensions his challenge induced and the impossibility of its ever succeeding, since (as he himself argued) it went against Nature.

> Nature is obstinate, and will not quit the field, however strongly attacked by reason; and at the same time reason is so clear in the point that there is no possibility of disguising her. Not being able to reconcile these two enemies, we endeavour to set ourselves at ease as much as possible . . .[45]

His philosophy can scarcely be regarded as a remedy then, he concedes; indeed, the skeptical doubt it encourages can just as well be thought 'a malady which can never be radically cured'.[46]

The price paid by this version of eliminative empiricism was, therefore, very high. It might be argued that the fault lay with the theory of ideas it presupposed, and not with the eliminativist program itself. But this is only partially true. If the danger incurred by realism has always been credulity or the acceptance of more than there is, the constant danger run by eliminativist proposals is skepticism, an undermining of belief even in what in fact there is. This is the great divide. The main purpose of this historical excursion has been to recall the contortions that eliminativism has so often in the past forced its proponents to engage in. This alone might be thought to constitute argument enough for some sort of realist alternative!

## 5 The realist thesis

Back finally to the realist thesis: what it is and what it isn't. Is it really a matter of 'explaining the success of science', as it has often been portrayed? There are at least two drawbacks about this way of presenting the issue. The first is that it makes realism sound like a *global* thesis about science in general. But there is no global realist implication. Some theories are more successful than others; some are longer-lasting than others; some, like geology, lend themselves more easily to a realist reading than do others, like mechanics. Besides, success can take many different forms, and the differences are very likely to affect the realist bearing of the theories concerned.

Although this form of global approach to the problem of realism is clearly objectionable, some questions about realism *can* legitimately be posed in a more or less general way. One can appropriately ask, for example, whether explanatory success of a particular kind can warrant some degree of belief in the truth of a theory (and hence in the existence of the entities it postulates as likely causes). Or again: if a great many of the theories that have become entrenched in the natural sciences over the last couple of centuries have enjoyed a striking degree of explanatory success (and sometimes a rather direct form of verification), might one not draw a cautiously realist conclusion about natural science in general, albeit with some strong qualifications? The danger here is that one may be tempted to forget that the question about the realist bearing of theory must be posed in the first place to the individual theory, even though the philosopher may want to generalize about the criteria that enable the individual judgement to be formed. It is the working *scientist*, not the philosopher, who utilizes such criteria when estimating the likelihood of a particular theory.

Taking realism to be primarily an explanation of 'the success of science' has a further, and potentially more troublesome, consequence. The impression is given that there are two separate levels of explanation involved in the defense of the realist thesis. First, the scientific theory explains the phenomena successfully; then, the realist thesis is held to be warranted by the fact that it explains the success of the theory. Critics have charged, not unreasonably, that this risks an obvious circularity.[47] If a question is raised as to whether explanatory success warrants (qualified) belief in the truth of a scientific theory, one cannot assume in responding that the explanatory success of the realist thesis, in accounting for the success of specific theories, warrants belief in its truth. It is the same form of argument in both cases; if it is challenged in one, its validity cannot convincingly be presupposed in the other as a way of legitimating the first.

The fault lies with the rather muddy notion of 'explaining the success of a theory'.[48] In order to see this, consider a concrete example, the plate-tectonic theory in geology. Postulating immense slow-moving tectonic plates has successfully explained a very wide range of geological phenomena, from mountain building to the gradual separation of continents. Does this warrant belief in the existence of these plates? The 'explanationist' answer is that it does, because only their existence would explain the success of the theory. But notice that there are not two separate explanations here. There is only one explanation, that involving tectonic plates. Their activities explain the varied geological phenomena in a highly successful way. And they may also be said to explain the historical fact that the plate-tectonic model has had the numerous successes it has had in explaining the phenomena. Are these really different explainers? Doesn't the second simply redescribe the first in a more complex second-order way? If the two explanatory factors reduce to one, is the charge of circularity negated?

There is not space here to deal with this tangled issue properly. But it may be worth sketching a possible response from the 'explanationist' standpoint. (In the final section of this essay, we shall explore the advantages of a rather different standpoint.) Back to the plate-tectonic theory: a critic might object to the claim that its success warrants belief that there are tectonic plates, on the grounds that another theory might perhaps explain the phenomena even more successfully. The defender of the realist construal of theory might then respond that this would still leave the previous successes of the first theory unexplained. The critic in this case has to attribute the prior explanatory success to accident, in other words to leave it unexplained. The persuasive power of the argument will depend on how significant the unexplained residue appears in the given case. Adopting the explanationist strategy means concentrating on the alternatives of explaining or leaving unexplained, rather than on defending the credentials of one theory against the possibility of a better one.

Logically the two are equivalent, but in rhetorical terms the explanationist strategy is clearly the more effective. It will not, of course, convince someone who, like van Fraassen, denies epistemic force to explanatory power in the first place. His objection to allowing explanatory success of itself to serve as warrant for the claim that tectonic plates exist is not the possibility of another better theory but the effective dismissal of explanatory power as anything other than pragmatic in the first place. This counts as directly against the second-level explanationist strategy as it does against the first-level realist construal of a theory. But, of course, it *does* require the sort of tolerance for the unexplained that empiricists of the eliminationist variety must possess in abundance.[49]

## 6 The reach of theory

And so, finally, we return to the question with which we began. What sort of assurance have we that the natural sciences can (as the realist claims) open up worlds that lie outside the narrow compass prescribed by the eliminative empiricist? One could begin here with the role played by instruments. That astute critic of empiricism, Francis Bacon, once remarked: 'The testimony and information of the sense has reference always to man, not to the universe; and it is a great error to assert that the sense is the measure of things'.[50] Bacon was writing at a time when the shift towards instrumentation had barely begun; his advice, nevertheless, was to rely on 'the subtlety of experiments [which] is far greater than that of the sense itself . . . the office of the sense shall be only to judge of the experiment; the experiment itself shall judge of the thing'.[51] He could never have anticipated the modern scientific laboratory where virtually none of the instruments are measuring properties that are accessible to the unaided human senses. Are these properties real? Even the hard-core empiricist would allow them some sort of dispositional or operational status. ('If you hook up this wire in this way, you will get the following reading on this instrument.') But the realist will want to name them as intrinsic properties ('electrical resistance') whose relationship to other properties may be reliably understood in theoretical terms. This first, and crucial, enlargement of the world is brought about by sophisticated instruments whose operation is interpreted theoretically, yielding a network of causally interconnected properties that lie outside, often very far outside, the narrow reach of the human senses.

Implicit in this enlargement is the notion of lawlikeness, of empirically discovered and theoretically interpreted regularity, akin in some (not all) respects to what Aristotle meant by *nature*. How can we tell that things have natures, that they can be counted on to behave in regular ways, that to the extent that a departure from regularity occurs, a further explanation for this can reasonably be sought? This is, of course, the familiar problem of induction in one of its guises. A realist might respond that postulating natures affords the best explanation for our immensely varied experience of the physical world, both in the everyday contexts supporting the Aristotelian account of nature and in the technical contexts of modern experimental science. The appeal here is to retroduction, not just to induction alone. Hume's demand for demonstrative proof is rejected as inappropriate. The knowledge of nature, grounded in this way, is sufficient to warrant rational action for the future. Obviously, more needs to be said but our main topic still lies ahead.

The principal way in which natural science enlarges our world is, as we have seen, through retroductive inference to structures, processes and entities postulated to be causally responsible for the regularities established by the experimental scientist, or for the individual 'traces' with which historical sciences like geology and evolutionary biology are concerned. They are *postulated* to be responsible; they are not observed, though they need not be unobservable. The warrant for their existence as causes lies, therefore, in the quality of the inference they permit, or more generally in the epistemic credentials of the theory resting on them. Everything rests, therefore, on these credentials. The realist force of any given theory depends in the first instance on how the theory itself is evaluated *as* a theory, on how good a theory it has shown itself to be. This is an everyday matter on the part of the scientists involved; it is not some sort of second-order judgement on the part of philosophers. Nor need it involve an explicitly realist emphasis ('do tectonic plates really exist?'); it suffices if it is a broad-based assessment of the scientific theory itself ('how strongly supported is the plate-tectonic theory?').

But what is still needed (and this is where the philosopher may contribute) is an analysis of the criteria involved in theory appraisal. How far can they go in assuring the truth of the theory and therefore the existence of the entities it postulates? The eliminative empiricist maintains that all that matters is empirical adequacy ('saving the phenomena'). As long as the theory predicts more or less correctly, further questions as to whether it is adequate as *explanation* bear only, in van Fraassen's view, on pragmatic issues of no relevance to the theory's epistemic status: 'They do not concern the relation between the theory and the world, but rather the use and usefulness of the theory; they provide reasons to prefer the theory independently of its truth'.[52] Furthermore, empirical adequacy itself furnishes, according to him, only grounds for what he calls 'acceptance' of a theory, not a warrant for believing it to be true. Since theories are to be evaluated only on the grounds of their empirical adequacy, they function only as a means of prediction; the question of their truth is not a properly scientific concern.[53]

Here the issue between realist and anti-realist is clearly joined. Might not the empirical adequacy of a theory alone have a positive bearing on its truth? Is it the case that the other virtues scientists look for in their theories are irrelevant (as anti-realists maintain) to the relationship between theory and world? What is the warrant for taking these virtues to *be* virtues of theory? These are large questions subject to much controversy; an outline response will have to suffice. First, what are these extra qualities that are regarded as marks of a good theory? Let us call them the 'complementary' virtues of theory, since they are assumed to complement the basic warrant given by a simple 'saving of the phenomena'.[54]

The importance of these virtues in helping to determine the truth of theory was a persistent theme in seventeenth-century accounts of scientific method.[55] Writers as philosophically diverse as Kepler, Descartes and Boyle, realized that hypothesis was required in order to reach causes which because of their minuteness or their distance from us lay outside the normal scope of induction. They also saw that the criterion of 'saving the phenomena' had to be supplemented somehow; it had long been evident in astronomy that quite different models might equally well save the phenomena. But what were these other criteria to be? Each had his own suggestions to make; once the simplicity and security of the traditional deductive method of demonstration were forsaken, it was not easy to find epistemic footing. Nor is it still.

For our purposes here, the candidates for the role of complementary virtue might conceivably be divided into four groups. There are internal virtues, such as logical consistency, coherence (or 'naturalness', absence of *ad hoc* features) and causal specificity. There are external virtues of consonance: consonance with other parts of science and (more controversially) consonance with broader world-views (with metaphysical principles of natural order, for instance). There are what might be called diachronic virtues, virtues that only reveal themselves over time as a theory develops and meets new challenges, such virtues as fertility and consilience (the unification of domains previously thought disparate). And finally, there is uniqueness, the absence of credible theoretical alternatives.

The reader may be struck by two omissions: simplicity and explanatory power. Simplicity is omitted because of its ambiguity as a requirement. In one meaning of the term, common among scientists, it comes close to what was termed coherence above; in another (favored by the logical positivists) it is a matter of practical convenience; in yet another it is an aesthetic quality trusted by some (e.g. Einstein and Dirac), distrusted by others (e.g. Nancy Cartwright), as an indicator of truth. Explanatory power (or explanatory success) is omitted from the list because it is too unspecific as a criterion. In particular, it does not contrast with the virtue of saving the phenomena, since a good explanation clearly has to get the phenomena right. Some of the complementary virtues bear only minimally, if at all, on the explanatory power of a theory, logical consistency and uniqueness, for example. When explanatory power is emphasized as a criterion, it usually refers to a *cluster* of virtues, notable among them coherence, causal specificity, and the diachronic virtues of fertility and consilience.[56] I shall use the term in this cluster sense.

The complementary virtues make sense only in a realistic perspective. This is the point to which we have been leading. There are many ways to make the case. It will have to suffice to outline just a few of these. Consider, for example, the role played by the virtue of *coherence* in major theory decisions of the past. The Copernican debate is as good an example as any.[57] Unlike the mathematical astronomers preceding him, Copernicus claimed that his model was not just an economic saving of the appearances but that it was *true*, so that the earth really *was* in motion. To appeal only to the appearances (to the criterion of empirical adequacy) would not suffice. His own system would have to be presented as a better *explanation*. And so he noted several well-known items of planetary lore that could be accounted for in an entirely 'natural' way in his system but appeared as *ad hoc* in Ptolemy's. The upper planets are at their brightest when in opposition to the sun, for example. This is a necessary consequence if the earth and outer planets move around the sun as center, and brightness depends on distance. But in a geocentric model, it has to be added in as a contingent – and puzzling – feature. Kepler developed this line of argument in much more detail later, noting in particular that in Ptolemy's epicyclic model of the planetary motions, the period of rotation on one of the two circles for each planet had to be *exactly* one year. But why, if each was, as Ptolemy supposed, an independently moving body? One could, of course, arbitrarily postulate this as a feature of the model, as Ptolemy did. But, inevitably, so striking a coincidence would cry out for an explanation, which only a system in which the earth itself was assigned a one-year period around the sun could give.

One could always claim, of course, that the planetary parameters just 'happened' to be that way. And as long as the arguments against the earth's motion seemed strong, as they did until the appearance of Galileo's *Dialogue*, this defense would have to suffice. But in the eyes of Copernicus and Kepler, the appeal to explanatory coherence was already powerful enough to overcome even the apparently decisive objections to the earth's motion that had blocked serious consideration of the heliocentric alternative for two millennia. This episode makes little sense from the strict empiricist perspective. If saving the phenomena is the only consideration that counts, then the arguments of Copernicus and Kepler ought to carry no weight. Yet we would want to say that even before Galileo produced new phenomena, the Copernican argument could already legitimately claim some weight.[58] It was not decisive, of course. It was *possible* that all of these coincidences were nothing *but* coincidences. That is the line the eliminative empiricist always seems forced to take.

Arguments of a similar sort might be built around other specific virtues, like fertility or consonance or uniqueness.[59] Some more general arguments may, however, be preferable in closing. What is the significance of the role played by the complementary virtues in general in theory assessment? First, there can be no dispute about the fact that scientists do rely heavily on these indicators, or that doing so has enabled them to shape theories that also saved the phenomena more and more effectively. Van Fraassen concedes: 'But it might be arguable that, for purely pragmatic (that is, person- and context-related) reasons, the pursuit of explanatory power is the best means to serve the central aims of science', the primary aim being 'to give us theories which are empirically adequate'.[60] But why *are* they the best means to serve this aim, if indeed they are purely pragmatic? What explains their effectiveness in promoting the central empiricist aim? The realist has a ready answer to this question; strict empiricists have to leave it unanswered; they have to regard it as a brute fact about how science operates.

The realist will, of course, deny that the explanatory virtues are entirely 'person- and context-related', in van Fraassen's phrase. The quest for causal explanation in terms of underlying structures and processes has been an invariant in the natural sciences since the seventeenth century.[61] The degree of emphasis laid on it has, as we have already noted, varied from science to science, from period to period, and even from scientist to scientist. When Kuhn drew attention to this feature, he went on immediately to add that 'roughly speaking' the basic virtues involved in theory choice have been 'permanent' features of theory-appraisal over recent centuries, 'unaffected by their participation in transitions from one theory to another'.[62] Retroductive argument from effect to postulated cause has been enormously successful, not least in leading to causes that subsequently have been observationally verified, whether the lunar mountains of Galileo's *Dialogue* or the large molecules of the contemporary biochemist. Reliance on the explanatory virtues has permitted a progressively finer detailing of the underlying structure of everything from cells to galaxies. What is impressive is the *continuity* of that effort; what makes each step possible is reliance on the assumption that the earlier steps were indeed warranted, that the structures proposed were in reality more or less as the theories said they were. This progressive elaboration of specific structures characterizes the history of much of natural science: chemistry, biology, geology. (Mechanics, as always, is a special case.) It is impossible to make sense of this if one is working within the limitations advocated by the strict empiricist.

The complementary virtues as a whole make sense in the realist perspective. They are just the norms that one would *want* to impose on theory if theory is understood as the realist understands it. The fact that they are a constant feature of scientific practice comes, then, as an implicit confirmation of that perspective and of the conviction prompted by it that causal theory, properly guided, gives reliable access to a world far wider than the narrow world of the empiricist from which scientific inquiry long ago began.

## Notes

1. The phrase is Bas van Fraassen's in the closing lines of his chapter, 'The world of empiricism'. Whether he can, as he hopes, distance himself from instrumentalism and still maintain this phrase as his motto seems, on the face of it, questionable. More of this later.

2. Sociologists of science are another matter. Though they differ strongly among themselves on matters of method, such critics of traditional notions of scientific objectivity as Steve Woolgar, Michael Lynch, and David Bloor seem to be advocating in practice something very like an instrumentalist thesis in regard to the claims of scientific theory.
3. For a historical sketch of the transition from the earlier model of science as demonstrative to the more relaxed notion of science as retroductive, see McMullin, *The Inference that Makes Science* (Milwaukee, Marquette University Press, 1992).
4. In 'The world of empiricism', van Fraassen focusses the discussion almost entirely on the question as to whether there are laws of Nature, or intrinsic 'necessities', involved in physical agency. He leaves retroduction aside entirely, yet what the reader wants insistently to know is whether the world of the empiricist contains such things as electrons, genes, and neutron stars. If so, on what grounds, if not explanatory success? If not, that world is far more constricted than the sole announced hesitation regarding necessities would lead one to expect. I shall return to these queries later.
5. Ian Hacking is skeptical of retroduction, but defends a form of 'experimental' realism that relies on manipulability as a criterion for discriminating between the real and the artifact in experimental science. He is especially critical of any form of empiricism that would calibrate the limits of acceptable existence claims by the bounds of the human senses (*Representing and Intervening* (Cambridge University Press, 1983)). Though retroduction is not the only strategy to which a defender of realism can point, it is still, however, the means of 'world enlargement' on which the ampliative claims of natural science primarily rest.
6. I avoid this expression here, preferring the Peircean term 'retroduction'. One does not, strictly speaking, infer *to* the best explanation; one infers that, in a particular context, a particular explanation *is* the best explanation. The distinction is an important one, for reasons made clear during the debates of yesteryear regarding a 'logic' of discovery.
7. Hilary Putnam, *Reason, Truth and History* (Cambridge University Press, 1981), p. 49; *Realism with a Human Face* (Harvard University Press, 1990), p. 30.
8. See, for example, Putnam, *Meaning and the Moral Sciences* (London, Routledge, 1978), pp. 123–40.
9. Putnam, *Reasons, Truth and History*, p. 54.
10. Although Putnam flatly rejects the claim that 'truth involves some sort of correspondence' that he associates with metaphysical realism, he does appear to allow a sort of correspondence, after all, as part of his internal realism: 'In an internalist view also, signs do not intrinsically correspond to objects, independently of how those signs are employed and by whom. But a sign that is actually employed in a particular way by a particular community of users can correspond to particular objects *within the conceptual scheme of those users*' (*Reason, Truth and History*, p. 52, his emphasis). It is hard to see how language could function without this

minimal sort of correspondence. And if this sort be admitted, it would seem that all he is rejecting is the 'similitude' notion of reference (*ibid*, p. 56) which would make our concepts *resemble* the objects they represent. But one can surely reject this, the classic Aristotelian theory of the concept, without rejecting the notion that truth involves *some* sort of correspondence.

11. Putnam, *Reason, Truth and History*, p. 57.
12. *Ibid*. pp. 52–5.
13. *Ibid*. p. 73. 'No independent access' arguments of this sort can be found also among critics of scientific realism. They depend, of course, on the Kantian premiss that things-in-themselves are entirely inaccessible. And they also suppose that independent access is required in order to make a case for correspondence (as well as for the existence-claims made by the scientific realist).
14. *Ibid*. p. 73. Newton did not, in fact, describe gravitation as acting across empty space. Indeed, in a letter to Richard Bentley he remarked: 'That gravity should be innate, inherent, and essential to matter, so that one body may act upon another at a distance in a vacuum without the mediation of anything else . . . is to me so great an absurdity that I believe no man, who has in philosophical matters a competent faculty of thinking, can ever fall into it.' (*Isaac Newton's Papers and Letters in Natural Philosophy*, ed. I. B. Cohen (Harvard University Press, 1958), pp. 302–3). Newton did not know what to make of gravity; in an earlier letter to Bentley written in 1693, six years, that is, after the appearance of the *Principia*, he had said: 'The cause of gravity is what I do not pretend to know and therefore would take more time to consider of it' (*ibid*. p. 298). And he *did* take a great deal of time to consider it in his later years (see McMullin, *Newton on Matter and Activity* (University of Notre Dame Press, 1978), chap. 4). His favourite speculation that there are 'active principles' operating upon bodies and somehow present in the spaces between bodies, was clearly closer to the later notion of a field than to that of action at a distance. This, of course, in no way weakens the ability of this example to make Putnam's point; Newton's puzzlement about how to assign the 'cause' of gravity reflects the multiplicity of incompatible causal interpretations of the same formalism.
15. See, for example, James Cushing, 'Historical contingency and theory selection in science', *PSA 1992*, ed. D. Hull *et al.* (E. Lansing MI, Philosophy of Science Association, 1992), 1, 446–57.
16. Richard Rorty, *Consequences of Pragmatism* (Minneapolis, University of Minnesota Press, 1982), p. xxi.
17. *Ibid*. p. xvi.
18. *Ibid*. p. xvii.
19. Richard Rorty, 'Is natural science a natural kind?', *Construction and Constraint*, ed. E. McMullin (University of Notre Dame Press, 1988), 49–74, esp. p. 58.
20. *Ibid*. pp. 49–51

21. In E. McMullin 'A case for scientific realism', *Scientific Realism*, ed. Jarrett Leplin (University of California Press, 1984), p. 24, I claimed that retroduction is 'central' to natural science. Rorty infers from this that I would also say that 'scientific explanation is of a distinctive sort – that science can be distinguished from non-science by its use of a special sort of inference' (Rorty, 'Is natural science a natural kind?', p. 59). On the contrary, I would argue that retroduction does *not* afford a principle of demarcation. Nor can I see why a realist should require it to do so.
22. Rorty remarks: 'It takes, after all, a good deal of acculturation to see the point of questions like: "Do numbers, or justice, or God, exist in the sense in which goldfish do?" Before we can get our students to approach these questions with appropriate respect, we have to inculcate a specifically philosophical use of the term "existence"' (*ibid.* p. 57). And he links that special sense with the 'invidious hierarchy' of the divided line and the distinction between primary and secondary qualities. (In a celebrated paper, Rudolf Carnap asked, 'What does acceptance of a kind of entities mean?' and made a similar sort of move, assimilating the question of the existence of the theoretical entities of the physicist to the question of the existence of numbers; see Rudolf Carnap, 'Empiricism, semantics, and ontology', in his *Meaning and Necessity* (University of Chicago Press, 1956), pp. 205–21. Asking about the 'existence' of numbers or propositions does, to be sure, require one to ask whether or not a special sense of 'existence' is needed. But this is not true for the theoretical entities of the natural scientist.
23. Rorty, 'Is natural science a natural kind?', p. 69.
24. As does van Fraassen's 'pragmatic' analysis of scientific explanation in *The Scientific Image* (Oxford, Clarendon Press, 1980), chap. 5. 'It might be thought that when we request a *scientific* explanation, the relevance of possible hypotheses, and also the context-class [the elements on which his pragmatic account rests] are automatically determined. But this is not so, for both the physician and the motor mechanic are asked for a scientific explanation [of a death in a particular car accident]' (p. 129). The examples on which van Fraassen relies all bear on particular events (like car accidents). It is precisely to eliminate the context-dependence of the explanations of such events that scientists devise the controlled environment of the laboratory. The explanations, both complementary and competing, offered for a contingent event like a car accident may make use of sciences borrowed from elsewhere, but do not contribute to an understanding of scientific explanation as this is found in physics or chemistry.
25. Thomas Kuhn, 'Objectivity, value-judgement, and theory choice', in *The Essential Tension* (University of Chicago Press, 1977), pp. 320–39.
26. See C. F. Delaney, 'Peirce on objectivity and truth in science', in *The Social Dimensions of Science*, ed. E. McMullin (University of Notre Dame Press, 1992), pp. 27–46.
27. See, for example, Carl G. Hempel, 'Valuation and objectivity in science', in

*Physics, Philosophy, and Psychoanalysis*, ed. R. S. Cohen and L. Laudan (Dordrecht, Reidel, 1983), pp. 73–100; E. McMullin, 'Values in science', *PSA 1982*, ed. P. Asquith and T. Nickles, 2 (E. Lansing, MI, Philosophy of Science Association, 1983), 3–25.

28. *Ibid.* p. 66.
29. E. McMullin, 'The goals of natural science', *Proceedings of the American Philosophical Association*, 58 (1984), 37–64. There are complex and debated issues here, but they lie outside the scope of the present paper.
30. Van Fraassen supposes Aquinas to favor indeterminism in this context ('The world of empiricism', section 2). A stone might thus 'just of itself' begin to roll down a hillside even though a science of heavy bodies might still for Aquinas be said to be of the necessary. I am skeptical of this interpretation. The Aristotelian maxim, 'whatever moves is moved by something other than itself', does not admit of exceptions; stones do not on occasion just of themselves start rolling. What is lacking to the chance event is not a determining efficient cause but a single *final* cause that would situate the event in a unified ordering to the good.
31. There are signs in Aquinas' writings of an awareness that coming to know the essences of physical things is more problematic than the straightforward Aristotelian doctrine would lead one to suppose. See E. McMullin, *The Inference that Makes Science*, pp. 35–40.
32. Later nominalists departed further from the Aristotelian conception of science than did Ockham. Nicholas of Autrecourt, for example, maintained that certainty can be attained only in propositions that reduce to the principle of non-contradiction. From the existence of one thing the existence of another can never be inferred with certainty. Causal generalizations are thus at best only probable. The theological roots of nominalist doctrine are particularly evident in Autrecourt's writings; his interest is, first and foremost, in sustaining God's absolute freedom and he cites the uncertainty of our best knowledge of the sensible world to urge the superiority of what religious faith offers.
33. 'The world of empiricism', section 3 (this volume).
34. See Andre Goddu, *The Physics of William of Ockham* (Leiden, Brill, 1984).
35. See James Farr, 'The way of hypotheses: Locke on method', *Journal of the History of Ideas, 48* (1987), 51–72; Larry Laudan, 'John Locke on hypotheses', in his *Science and Hypothesis* (Dordrecht, Reidel, 1981), 59–71; E. McMullin, 'Conceptions of science in the Scientific Revolution', in *Reappraisals of the Scientific Revolution*, ed. David Lindberg and Robert Westman (Cambridge University Press, 1990), 27–92 (especially pp. 75–6).
36. John Locke, *Essay Concerning Human Understanding*, Book IV, chap. 16: 'Degrees of assent'. He also speculates that a more direct access may some day become available through the use of powerful microscopes that could magnify 'a thousand or ten thousand times' (*ibid.* Book II, chap. 23, para. 11). He has no problem about treating this as simply an 'augmentation' of our sense knowledge.

37. From a brief manuscript titled 'Method', reprinted in Farr, 'The way of hypotheses', pp. 70–2.
38. See note 14 above.
39. George Berkeley, *De Motu* (1721), see *The Works of George Berkeley*, ed. A. A. Luce and T. E. Jessup (Edinburgh, Nelson, 1952), vol. 4, para. 17. He had already announced this theme in *The Principles of Human Knowledge* (1710), see *Works*, vol. 2, paras. 103–10.
40. *Ibid.* paras. 37, 71.
41. *Ibid.* para. 35.
42. *Ibid.* paras. 37, 71.
43. See, for example, Margaret Wilson, 'Berkeley and the essences of the corpuscularians', in *Essays on Berkeley*, ed. J. Foster and H. Robinson (Oxford, Clarendon Press, 1985), pp. 131–48; J. O. Urmson, 'Two central issues in Bishop Berkeley's corpuscularian philosophy in the *Siris', History of European Ideas*, 7 (1988), 633–41.
44. See Barry Stroud, *Hume* (London, Routledge, 1977), chap. 10.
45. David Hume, *A Treatise of Human Nature*, ed. L. A. Selby-Bigge (Oxford University Press, 1958), p. 215.
46. *Ibid.* p. 218. See Stroud, *Hume*, chap. 5.
47. Laudan calls it 'the ultimate *petitio principii*' in 'A confutation of convergent realism', in *Scientific Realism*, ed. Jarrett Leplin (University of California Press, 1984), 218–49, p. 242. See also Arthur Fine, 'The natural ontological attitude', same volume, 83–107.
48. McMullin, 'Selective anti-realism', *Philosophical Studies*, 61 (1991), 97–108.
49. In his account of explanation, van Fraassen emphasizes the importance of what he calls 'rejection', where the request for explanation is legitimately denied, either because it is not needed (uniform motion; principle of inertia), not possible (the time of decay of a single radioactive atom; quantum mechanics), or not yet available (why one person with syphilis contracts paresis as against others with the same condition) (*The Scientific Image*, pp. 111–12). But in each of these cases one can give an account of why the demand for explanation is denied or postponed. According to the realist, that is precisely what cannot be done in the matter of explaining the long-term success of specific scientific theories.
50. Francis Bacon, *The New Organon*, ed. Fulton Anderson (Indianapolis, Bobbs Merrill, 1960), p. 21.
51. *Ibid.* p. 22.
52. Van Fraassen, *The Scientific Image*, p. 88; see also p. 156.
53. *Ibid.* p. 10. This may seem a distinction without a difference, however, if both sides agree in their main contention that the function of theories in science is to serve simply as instruments of prediction and economic organization of data.
54. Paul Churchland speaks of 'superempirical' virtues ('The ontological status of observables: in praise of the superempirical virtues', in *Images of Science*, ed. P. Churchland and C. Hooker (University of Chicago Press, 1985), 35–47). Richard Boyd prefers the term 'non-experimental'

('Observations, explanatory power, and simplicity; towards a non-Humean account', in *Observation, Experiment, and Hypothesis in Modern Physical Science*, ed. P. Achinstein and O. Hannaway, (Cambridge MA, MIT Press, 1985), 47–94). Neither of these terms is altogether adequate; each has misleading associations which the more neutral term, 'complementary', avoids.

55. McMullin, 'Conceptions of science in the Scientific Revolution', in *Reappraisals of the Scientific Revolution*, ed. D. Lindberg and R. Westman (Cambridge University Press, 1990), 27–92.
56. McMullin, 'Values in science'. The list given in this earlier essay differs in several respects from the one outlined above. A much fuller account is undertaken in 'Realism and epistemic virtue', in preparation.
57. The Copernican story is told at much more length in McMullin, 'Rationality and paradigm change in science', in *World Changes; Thomas Kuhn and the Nature of Science*, ed. P. Horwich (Cambridge MA, MIT Press, 1993), 55–78, pp. 71–5.
58. I have deliberately left an ambiguity in this account. Is the realist appealing to the historical fact that Copernicus and Kepler used an argument of this sort, one that later was so spectacularly vindicated? Or that it was a *good* argument? Or both?
59. Whewell built his well-known argument for realism on *consilience*, which, in favorable cases, gives to 'theory a stamp of truth beyond the power of ingenuity to counterfeit' (*The Philosophy of the Inductive Sciences* (London, Parker, second edition 1847), vol. 2, p. 66). Several recent writers (notably Michael Friedman and Philip Kitcher) have explored the idea of identifying explanation in science with *unification*. Something must, however, be added to this: unification alone is clearly insufficient, as the example of Ptolemaic astronomy illustrates. One must unify disparate domains (make them 'jump together', as Whewell put it) *under a single causal story* for it to count as explanation in anything other than the weak Humean sense enshrined in the DN model of explanation. Consilience testifies to the truth of the causal story, though, of course, the assessment of that story as explanation will involve other virtues also. Unification of itself can easily be dismissed by the strict empiricist as a pragmatic virtue of convenience.
60. Van Fraassen, *The Scientific Image*, pp. 89, 12.
61. It is no accident that, from Berkeley's day to our own, anti-realism has flourished most among those who take mechanics as their paradigm of natural science. Van Fraassen has frequently chided realists on their maintaining a thesis which would compel them to defend 'hidden variables' in quantum mechanics. But the realist thesis is *not* a normative one in this sense; it does not compel one to hold that an underlying causal structure will be found in *all* cases. Rather, it is *retrospective*: it extracts a realist implication in those cases where the postulation of underlying causal structures has, in fact, proved successful.
62. Thomas Kuhn, 'Objectivity, value judgement, and theory choice', p. 335. In Kuhn's view, the virtues guiding theory choice are only *partially* pragmatic;

it is this that allows him to deny that he is depriving science of objectivity. Were he to push this point further, it might have led him to reconsider the anti-realism of the Postscript to his *The Structure of Scientific Revolutions*. See McMullin, 'Rationality and paradigm change in science'.

# 5 The world of empiricism

Bas C. van Fraassen

My topics are the relation between science and myth, and the possibility of empiricism as an approach to life as well as to science. But philosophy is a thoroughly historical enterprise, a dialogue that continues in the present but is always almost entirely shaped by our past. So I will devote the first half of this chapter to setting the historical stage.

## 1 Two philosophical traditions

There are two main traditions in philosophy about science and about our knowledge of nature; I'll refer to them as *realist metaphysics* and *empiricism*. Both of these can be approached more narrowly as concerning how we should understand science. But we can also think of them more broadly as concerned with making sense of the world and, at the same time, of our attempts to make sense of the world.

For the realist, science is a journey of discovery. In fact, realists think of philosophy and science as jointly trying to uncover what is really going on in nature, even 'behind the scenes' so to speak. And at the same time, the realist sees science as aiming at real understanding of how nature works, and why it is the way it is. The two aims, of discovering what the world is like, and understanding or making sense of it, are not automatically the same! But for realism there is no tension between them – they are happy to identify the What and the Why.

All of this presupposes of course that there is a *Why*, that for everything there is a reason. Another way to characterize realist metaphysics, more or less equivalently, is as follows. For realists the question *Why*? has absolute primacy; and they presuppose that it must have an answer always. Science is then conceived of as taking on the task of answering that question. What sort of answers is it supposed to get? Realist metaphysical systems (which are proposed as extensions of and continuous with science) – or extant science itself (under some realist interpretation) – give the answers, and do so by postulating 'deep' facts about the world. In other words, realists are satisfied with answers-by-postulate.

It is not equally easy to characterize empiricism. Mostly empiricists have distinguished themselves by their negative, critical reactions to various sorts of realism. Happily we possess a dramatization of what it is like to become an empiricist through bone and marrow. I at least have always understood Jean-Paul Sartre's novel *Nausea* as doing exactly that: the protagonist Antoine Roquentin is in the agonizing process of becoming what I'll call a *classic empiricist*. Eventually Roquentin says: 'Now I knew: things are entirely what they appear to be – and behind them . . . there is nothing.' This is an extreme form of empiricism: Roquentin denies the realist assumption that there is something to be known or found 'behind' the phenomena which appear to us. Later I will quote him further to illustrate some of his agony with this way of being in the world. I will also argue that not every way of being an empiricist needs to entail such agony.

Not every empiricist is equally extreme in his or her conclusions. To reject realism it is enough to become agnostic about what the realist says we need to find. But a thoroughgoing agnosticism may not be any more comfortable, for it lacks even the sense of certainty that comes with saying 'There is nothing there!'

The philosophical reactions that I identify as empiricist through the course of history have always in the first instance been rebellions against realism. The empiricist comes across as being 'against theory', calling us back to experience. He or she is thoroughly sceptical of the philosophical stories about why experience must be this way or that. Realists counter that this will lead us into utter, debilitating scepticism, that it will deprive life and the world of all meaning and intelligibility.

It is a sad fact that when empiricists have tried to do something more constructive, they have often just ended up doing metaphysics too. Often they were seduced by the assumption that we can give meaning to life simply by attributing some postulated 'deep structure' either to the world we live in, or to our experience. Empiricism cannot simply go at right angles to the realist course; if it is to work at all, it must step out of that plane of motion altogether. The realist sees our epistemic enterprise as achieving a world-picture, something that purports to be the 'One True Story of the World'. Outright denial of that view would push us simply into a rival world-picture. I want to raise the possibility of life without a world-view – at least without the sort of world-view that metaphysical realists hold out for us as the aim of science and philosophy.

## 2 Classical empiricism

The story I shall tell in this part is a drama in three acts. Aristotle insisted that science aims not just to describe the phenomena but to explain them. He then immediately went on to identify explanation with description of something 'deeper'. This led to a view of science as describing necessity in nature (as opposed to the 'merely' actual), or laws of nature (as opposed to 'mere' regularities). The nominalist/empiricist rebellion of the late Middle Ages challenged any such enterprise which requires empirical science to reach for something far beyond empirical ken. In the third act the realists face the empiricists with a tragic dilemma: either you resign yourselves to living in an utterly meaningless world, or you must believe something not because your experience leads you to it, but purely to escape this meaninglessness.

Of course I admit what will already be obvious to you from this little synopsis: I am giving you a rather biased history of philosophy. But you know my bias: I want to be an empiricist, in some way that makes sense for us today.

### 2.1 Aristotle's view of science: the What and the Why

Let us begin with Aristotle's account of the eclipse. We find him focussing on eclipses of the moon. Looking back from the twentieth century, we think immediately that the reason is not far to seek. A lunar eclipse is visible at the same time over a large part of the earth, while the solar eclipse is visible only in a small part. Therefore quite accurate predictions were possible for lunar eclipses, but not for eclipses of the sun. But when we then check Aristotle's discussions, we find to our surprise that they are not at all concerned with this difference in predictability!

What does he discuss then? He discusses what an eclipse *is*. To him, the aim of science is to reach understanding, to know the reasons why things happen the way they do. Then it turns out that according to him, we understand such a phenomenon as the eclipse when we know what an eclipse is:

> The question 'What is eclipse?' and its answer 'The privation of the moon's light by the interposition of the earth' are identical with the question . . . 'Why does the moon suffer eclipse?' and the reply 'Because of the failure of light through the earth's shutting it out'.
> (Aristotle, *Posterior Analytics* II, 2, 90, 7–18)

Persuasive. But he must have something quite special in mind when he emphasizes the 'is' in 'what is'. Eclipses are many things – perhaps, for example, the eclipse is the one lunar phenomenon that has struck terror in the hearts of millions – but they don't all help to tell us why there are eclipses. So Aristotle envisages a sort of hierarchy or priority of properties: some properties are *essential*, others merely *accidental*. The essential ones answer the Why-question as well.

This hierarchy in what the thing is, comes from an asymmetry in explanation. If A explains B, you cannot also say that B explains A. To take a modern example: The light reaching us from distant galaxies exhibits a red shift if those galaxies are receding from us; and vice versa: those galaxies are receding from us if their light has this red shift. Yet it is the receding motion which explains the red shift – and not vice versa. Aristotle himself gave two examples: the planets do not twinkle (unlike the stars) because they are near; the moon waxes and wanes as it does because it is spherical. In each case, we are disinclined to add 'and vice versa'.

To account for these asymmetries, then, Aristotle holds that some properties of a thing are essential to it, and others not essential but accidental. The paradigm of explanation for him, is appeal to what is essential in order to account for the accidental.

*But what is essential?* We are never told completely. Part of the answer is that *only what is necessary is essential*. Hence, an explanation of the phenomena through appeal to what is essential, is partly this: to show why the phenomena *had to be* what in fact they were, they *had to happen* in the way they did.

This is how the idea of necessity enters the discussion: the asymmetry of *Why*? engenders an asymmetry in the *What*? which is traced to an asymmetry between contingency and necessity. But this new distinction is no less mysterious than the preceding one.

### 2.2 Mediaeval realism

Aristotle's account of the world was developed in depth by the mediaeval philosophers, such as Thomas Aquinas and Duns Scotus. Because they were also theologians, they had in principle available a quite simple account of those distinctions which Aristotle had introduced into natural philosophy.

Let me first say something about how they saw the world. Imagine I am holding a piece of chalk. It must fall if I release it. It must break if I apply a mere ten pounds of pressure on the middle. Looking at it, you would see that it is white; but you would know that if I kept it hidden and merely told you that it was pure chalk. The reason – the mediaeval philosophers would say – is that all those things are necessary to chalk, it is the nature of chalk. To this extent the world is determined.

Besides this deterministic aspect of the world they also recognized *chance* and *free will*. In fact, we today can see a rudimentary idea of statistical science in what Aquinas says about this:

> The majority of men follow their passions, which are movements of the sensitive appetite, in which movements heavenly bodies can co-operate, but [a] few are wise enough to resist these passions. Consequently, astrologers are able to foretell the truth in the majority of cases, especially in a general way. But not in particular cases.
>
> Thomas Aquinas *Summa Theologiae* Ia, 115, 4, ad 3

The mediaevals saw much more chance, and much less determination in the world than people do today, even now.

But Aquinas, following Aristotle, would not have classed all that as science. A stone could begin to roll down a hillside 'just by itself', or a person could 'just decide' to give in to some passion. Such accidental phenomena are classed as not within the realm of scientific knowledge. The aspect of the world covered by scientific explanation was exactly what is determined by the natures of things, that is, what is necessary. Their world is partly indeterministic, but science describes what is necessary, i.e. deterministic about that world (see e.g. Aquinas, *Commentary on the Posterior Analytics*, I, 16, 6–7; *Summa Contra Gentiles* II, 23, 2).

I said that as theologians, they had an explanation available: some things God decreed, and some He left open (so to speak). Actually that just pushes the issue one step farther back. Here is the puzzle. Suppose – to take one of their examples – that wood always burns when heated. What exactly did God decree: that wood will always burn when heated, or that wood must always burn when heated? That certain things will always happen or always be thus or so, or that they *have to be* always that way? Are there some things that happen every time, but are not necessary?

Let me try to convince you that philosophers are not perverse when they draw this distinction. You make that distinction too. Suppose we could equate: $X$ never happens = $X$ is impossible. That would also mean: if something is possible, then it will happen at some time or other. If you believe that, then you never need to use the words 'possible' again, nor its cognates like 'impossible' or 'necessary', because, if you believe that, the things that are possible are exactly the things that are actual at some time or other. In that case you could always use the word 'sometimes' instead of 'possible'.

But I'm sure you are not happy with that idea. For example, you may believe that the human race will never enter a giant suicide pact and do away with itself – but you won't say that it couldn't possibly do so. Even what you and I could do today, what is possible for us, is not exhausted by the few things we will actually do at some time or other.

Why is this important? Well, if those mediaeval realists are right about what science is, then science has to discover not just what always happens, but what is necessary. In other words, the task of science is to divide the things that always happen in a regular way into two classes: the ones that just happen to be that regular ('mere' regularity), and the ones that are that regular because that is how they have to be ('necessities in nature', 'laws'). But the problem for us humans is that we can't tell the difference just by looking at what actually happens – and still, that is all that we ever get to see.

How would it help to bring God into the picture? If the distinction is one He can make, if it comes for example from what He decreed and left open at Creation, we still have to ask: how can we make sense of science as a humanly possible enterprise? Aquinas himself insisted on making room for this enterprise, though with the reason that, to think of every necessity as directly connected to God's will (rather than having been instilled in nature, so to speak) was to denigrate Creation.

### 2.3 The nominalist/empiricist rebellion

Happily I can refer you to another literary dramatization: Umberto Eco's *The Name of the Rose*. The character William of Baskerville personifies the intellectual rebellion of such philosophers as William of Ockham. The nominalists of the late Middle Ages concluded there are no necessities in nature, no necessary connections (in later terminology: no laws of nature). On the theological side they merely said that of course it was necessary that if God willed something then it happened. But within nature itself, there is no division between necessary and merely actual.[1]

These nominalists of the fourteenth century turned upside down the whole conception of the world and of science. Their movement marks the true birth of empiricism. Of course they also have to show us that they can make sense of the world we live in, and of the scientific enterprise. Typically they begin with a simple point from logic. Consider the assertion that wood must burn when heated. Here is a piece of wood, namely a table. Should we conclude that it will necessarily burn if heated? Is it impossible for it to stay whole if I put it in a furnace? You are probably nodding *yes* to yourself.

But let me give you a parallel example. It is not an accident or coincidence that all bachelors are unmarried. Bachelors are necessarily unmarried. Now let me see if there is one in the audience. You, sir, are a bachelor? All right, then, should I now turn to you all and say: Behold, here we have one, a person who is necessarily unmarried, a person who could not be married? Actually there is nothing wrong with him, as far as I can see!

Similarly for a piece of wood: if it didn't burn when heated, we would say: It looks like wood in almost every respect, but it is not wood. If I say *this very thing* must burn when heated, that is elliptical – I expect the listener, to supply a tacit clause like 'given that it is made of wood, as we all know'. But in that case, the necessity is merely verbal.[2]

As you can imagine, this raised a storm of protest. To the realist it is crucial that some things are really necessary and really possible, not just verbally. You really cannot jump over a building, though you really can jump over a doorstep – that is not just a matter of words. Just try and you'll see! Don't such examples show that there are real necessities in nature? No, in fact they don't. For perhaps we just say such things – use the words 'really can, really cannot' – to express a very strong conviction that they won't happen. In that case, the use of those words says a lot more about us than about nature.

The realists answered that there were two things that all science aims at, and that are impossible for the nominalist. The first is *reasonable expectation*, and the second *explanation*. If there is no reason 'in' this table that *makes* it burn when heated, then there is no reason to expect that it will. No reason to expect that individual matches will burn, or water quench thirst next time you drink, or that Rosemary's baby will be human. And secondly, if nominalists are so perverse as to keep expecting that babies born from humans are always human, and so forth, then they have no explanation of that fact; the can't explain why the world should be so regular.

Let me quickly illustrate this with one of their actual disputes, which we look back on today as prefiguring Newton's first law of motion. When the thrown stone leaves the hand, what keeps it going? By Aristotelian principles either this stone continues to act on the air, cleaving it, or the air acts on the stone so as to push it forward. Neither sort of account was very successful; yet it was insisted that there must be some such reason. William of Ockham's reaction was characteristically radical: the question of what *keeps* the stone moving, he rejects.

> If you insist that the moving body does not move unless it acquires something which it did not have before, I answer that indeed it has

> something new, . . . namely a different location. And if you further inquire as to what is necessary for the body to be in that place, I reply that nothing else is required but a body and a place and the absence of any intermediary . . .'[3]

We probably shouldn't blame the mediaeval realists too much for resisting this. Even Newton, though he made it his first principle that things retain the same motion unless interfered with, still kept thinking that perhaps he had to appeal to some special sort of force (*vis insitae*, or *vis inertiae*) to make it intelligible. The realist instinct, that there must always be some deeper reason for everything, dies very hard.

### 2.4. The recent aftermath of this debate/Today

This debate, as I have said, began in the fourteenth century; these debates happened over five hundred years ago – but don't think that they aren't happening now! Writers who discuss today's physics sometimes have little sense of the history, and so they just repeat it, badly. The Einstein–Podolsky–Rosen paradox and violations of Bell's inequalities have furnished many examples of this, both from philosophers and from scientists.

A nominalist or empiricist does have to explain how reasonable expectation is possible. What such a philosopher must say is really quite simple: I believe, just as you do, that every time a stone is released it will fall. I believe that there is this regularity in the world. Whatever reasons you have for saying that something is necessary, are for me simply reasons to think that it is so.

The realist then says: OK, perhaps you can have expectations and predictions; however, you cannot explain them. You have no reasons to show why these things *have to happen* the way they do. The nominalist admits that, but does not give such primacy to the *Why*? question. There is an explanation, the nominalist says, but only in the mundane sense that sometimes we are puzzled, and we need the missing pieces for our puzzle – however very ordinary information will do just fine there, and science can serve us by providing this very ordinary sort of information about how things just happen to be.

In every century the battle of empiricism against realism is fought again. I don't expect you to be convinced yet by my side of the story.

Perhaps you too feel a great dismay that empiricism deprives us of so much that might comfort us in a hostile world. And it is true, it does: all it can offer is the agony and the ecstasy of freedom in a world governed by no laws except those we create ourselves:

> We are born by accident into a purely random universe. Our lives are determined by entirely fortuitous combinations of genes. Whatever happens happens by chance. The concepts of cause and effect are fallacies. There are only *seeming* causes leading to *apparent* effects. Since nothing truly follows from anything else, we swim each day through seas of chaos, and nothing is predictable, not even the events of the very next instance.
> Do you believe that?
> If you do, I pity you, because yours must be a bleak and terrifying and comfortless life.[4]

I quote this from Robert Silverberg's *The Stochastic Man*, a science fiction novel. This *is* the world of empiricism. It is the world of Sartre's hero Antoine Roquentin in *Nausea*, it is a world in which anything is possible, and whatever happens merely happens, and not because something greater is making it happen. Here is the famous passage in which those apparent limits to possibility dissolve before his eyes:

> I went to the window and glanced out . . . I murmured: *Anything* can happen, *anything*.

> Frightened, I looked at these unstable beings which, in an hour, in a minute, were perhaps going to crumble; yes, I was there, living in the midst of these books full of knowledge describing the immutable forms of the animal species, explaining that the right quantity of energy is kept integral in the universe; I was there, standing in front of a window whose panes had a definite refraction index. But what feeble barriers! I suppose it is out of laziness that the world is the same day after day. Today it seemed to want to change . . . then, *anything, anything* could happen.[5]

Roquentin also describes the security of others, who live in an illusory sense of ontological security:

> They aren't afraid, they feel at home. All they have ever seen is trained water running from taps, light which fills bulbs when you turn on the switch. . . .

> They have proof, a hundred times a day, that everything happens mechanically, that the world obeys fixed, unchangeable laws. In a vacuum all bodies fall at the same rate . . ., the public park is closed at 4 pm in winter, at 6 pm in summer, lead melts at 335 °C, the last street-car leaves the Hotel de Ville at 11:05 pm. They are peaceful, a little morose . . . Idiots.[6]

This is frightening; to lose our sense of necessity is to lose our sense of security.

But the danger of losing our emotional and intellectual comforts is not an argument. You will be reminded of those nineteenth-century clergy in Ibsen settings, losing their faith and arguing that religion was indispensable, because otherwise life would lose all its meaning, and they would not be able to continue to live. Well, as a philosopher I have to counsel suicide before an invalid argument.

So I conclude. There are no necessary connections in nature, no laws of nature, no real natural bounds on possibility. Those ideas all resulted when philosophers projected familiar models onto the natural world. Really, nothing is necessary, and everything is possible.

I mean this. All of the above is true. Yet I am not simply trying to persuade you that we have a bleak and comfortless life. What I reject is those philosophical ideas about where to turn for comfort. I am referring here to the realists' identification of understanding with knowledge of 'deep' facts about a reality behind the scenes of the phenomena. Science is our paradigm enterprise of empirical inquiry, and I value it very highly – but not as the acquisition of *such* knowledge. Now I had better try to make good on this by showing that there is another way to go.

## 3 Points of view/science and myth

Is there a constructive side to empiricism? Or does it make the search for meaning and meaningfulness hopeless? Is meaning just a matter of the psychopathology of everyday life?

### 3.1 What is our relationship with our world-pictures?

There is more to the role of science in our lives than prediction, expectation, and practical opinion. Science has transformed our world-view.

Empiricists have often been tempted by some form of instrumentalism: science is 'merely' an instrument. If science were a mere instrument, like an abacus or a calculator, how could it transform anything? Abacuses do not transform a world-view. Getting the idea of *how the abacus works* might do that, but not its mere use.

In attempting a positive account, I shall take a cue from Nietzsche, and liken science to myth. Myths, after all, do have the power to transform our consciousness of the world.

As soon as I say this, I know you begin to suspect the worst about me. After all, we use the word 'myth' in practice as a synonym for 'falsehood', and now maybe I am going to say that science is nothing but a myth. Well, I had better correct those impressions right away.

First of all, the word 'myth' does not strictly speaking imply falsehood at all. A Christian or Jewish theologian can certainly compare the Judaeo-Christian mythology with such rivals as the Pagan myths or the Hindu–Buddhist mythology. He or she does not say that these are all on a par, but only classifies what the Judaeo-Christian tradition gave us as significantly similar to those rival mythologies.

Second, about the 'nothing but' manoeuvre. This has an absolutely fatal fascination for philosophers. But 'nothing but' is logically not simple. Consider the statement: 'Jesus was nothing but a story-teller.' This presupposes that Jesus was a story-teller and then adds that he did not belong to any significant sub-class of story-tellers. To deny it categorically is to say that Jesus was indeed a story-teller, but of a special sort. After this preamble, I think you will not misunderstand me: I categorically deny that science is nothing but a myth.

### 3.2 What exactly is a myth?

Some discussions of myth by philosophers, literary critics, psychologists, anthropologists, and theologians, have aimed at a definition of myth. There is so little agreement, that I shall only try to describe salient features.

A myth is a story. This will have to be qualified; but let it stand for now. Myths must be distinguished from such types of stories as legends, parables, allegories, and popular history. Legends and popular history are stories that *purport to be true*. There is no such purport in the case of parables or allegories. These are distinguished by having a point or significance for morality or the meaning of life.

Sometimes myth is just defined as a combination of these two features: a myth is a story that both purports to be true and has the kind of significance that parables and allegories have.

This cannot be a good definition, for it would even make the story of George Washington and the cherry tree a myth. (Or the story, which is unknown in Holland but familiar to many foreigners, of the boy who saved Holland by putting his finger in the dike, to keep the sea out. In America they even know the name of that boy!)

Under the urging of anthropologists such as Malinowski, myths have now come to be discussed largely in terms of their function. In the nineteenth century there was a school that saw myths purely as embryonic science: the function of myth is to explain natural phenomena (paradigm: the Creation myths). The opposite extreme was espoused by philosophers like Ernst Cassirer and Susanne Langer. According to them, the function of myth was that it furthered a sense of harmony within society and with nature (paradigm: the Myth of the State). Today's anthropologists blend such extreme conceptions, and assign functions of both sorts to myth:

> The myths of the Australian aborigines, which deal with the creation of their Universe and the establishment of their rules of human behavior that all must follow, . . . are the foundations of their social and secular and ceremonial life.
>
> The myths that support these philosophies provide the aborigines with a reasonable explanation of the world in which they live; the stars above them, the natural forces of wind, rain, and thunder and the plants and creatures that provide them with food.[7]

I think there can be no doubt that science can and does serve functions of both those sorts. In his chapter, 'Quantum theory and our view of the world', Professor Feyerabend gives some examples of those social functions of science.

### 3.3 The rivalry between myth and science

One conclusion seems inescapable. Science presents itself, in each culture, as a rival to the mythical world-picture, and aims to replace it with a new world-view. To illustrate this, let me quote not a scientist but a theologian. Rudolf Bultmann emphasized this in the strongest terms in recent theological debates:

> The cosmology of the New Testament is essentially mythical in character. The world is viewed as a three-storied structure, with the earth in the centre, the heaven above, and the underworld beneath.
>
> All this is the language of mythology, and the origin of the various themes can be easily traced . . . To this extent *the kerygma is incredible to modern man, for he is convinced that the mythical view of the world is obsolete.*[8]

Bultmann adds, 'It is simply the cosmology of a pre-scientific age' and 'We no longer believe in the three-storied universe which the creeds take for granted.'

But where exactly does the rivalry lie?

There is no obvious rivalry between any little scientific theory and any little myth, such as, say, Archimedes' statics and the Oedipus myth. But the Greek mythology as a whole, like the Judaeo-Christian mythology and the Hindu–Buddhist or Islamic mythology, is a different matter.

So let us distinguish between *little myths*, like the Oedipus myth, and *great myths*, like the Judaeo-Christian, or the Hindu–Buddhist myth. This is the point where I must qualify the idea that a myth is a story. Certainly little myths are stories. Sometimes they are stories of what happened on specific occasions, such as the Fall, or Zeus' advent to hegemony; sometimes accounts of repeating or repeatable events, such as Apollo driving his chariot across the heavens from East to West every day, and the transubstantiation of bread and wine in communion.

Little myths are stories, and they are stories that change. Sometimes they die altogether, sometimes they re-emerge at a later point. A little myth may be born and die, in many versions that differ with time and locale, subject to different interpretations – and all that under the aegis of a single great myth. Little myths change, but the great myth endures.

The great myth too changes and develops. But its developments are not mainly changes in its narrative. The Judaeo-Christian myth had short periods of drastic change or rapid development around the times of Moses, St Paul, and Aquinas, to name but three. These were developments where we can truly speak of *conceptual revolutions*. At those points, there is not merely a change in little myths. For a long time, for example, God was an agent within history and within time. But at least as of the Middle Ages, God is trans-temporal and trans-phenomenal, trans-historical (in Jewish as well as Christian and Muslim theology).

Little scientific theories, like little myths, come and go: phlogiston emission gives way to oxidation, light particles to light waves to photons. Some little theories persist, but their details and the way they are understood changes from epoch to epoch: this is the way in which Archimedean statics and Huygens' theory of collision persist to this very day. Meanwhile, science endures: we are engaged in the same enterprise as Archimedes was.

But though it has endured, science has gone through several short periods of intense development amounting to genuine conceptual change, such as the Galileo–Newton and Planck–Heisenberg periods.

It is exactly here, in what we may call Great Science and Great Myth, where the main rivalry occurs, that we also see the most striking parallelism.

*Myth is cosmological*, presenting a picture which embraces the whole world and all of history. The drama it presents is an on-going one. *Science is cosmological*, global in compass, embracing the whole world and total world history.

*Myth is narrative*, in that it presents a drama unfolding in time, and a description of certain kinds of processes. But there is a point which, as far as I know, each great myth reached, at which the *dramatis personae* come to be seen as *transcendent*.

This is the *point of paradox*, where myth breaches the categories derived from common sense, and history is seen as a reflection of something not itself set in time and space. *Science too is narrative*, writing for us even a brief history of time – and of the cosmos evolving in time. But with relativity and quantum theory, it also reached the point of paradox. The categories of time and space are subsumed, '*aufgehoben*', in space–time; particles no longer have definite spatial trajectories; even duration and dates become subject to indeterminacy. The *dramatis personae* have become extramundane.

*Myth is explanatory*; it explains both the natural order and the development of the social order. *So does science.* Myth has a strong grip on the human imagination; it supplies the classification and the categories, the pigeon-holes and concepts, the *categorial framework* within which every subject is placed and understood. *So does science.*

### 3.4 Parallel debates concerning language

I just want to describe very briefly one more parallel. This concerns the irreducibility of the language of science and of myth, the impossibility of translating them into more 'hygienic' language.

Let me begin with the debate in recent theology, from which I have already quoted above, about demythologizing the gospel. Bultmann began by describing the mythical world-view that underlies the New Testament and contrasting it with our present world-view which is largely determined by science. He maintained that the gospel has not lost its significance, but that it needs to be represented in terms compatible with our present world-view.

After a great deal of demolishing, using both the scientific outlook and the results of historical scholarship, Bultmann also sketches a contemporary presentation of the gospel; and he does so in Existentialist terms.

But as Karl Jaspers pointed out, this is not demythologizing so much as translation into another, modern, mythical framework.[9] Perhaps the content of a myth cannot be rendered except in mythical terms. If so, myth is untranslatable, in a certain sense. We must distinguish this immediately from the sense in which poetry is said to be untranslatable. When a myth is translated into another language, even very badly as by a Sunday school teacher, it is still immediately felt as myth. This is also true of science, *mutatis mutandis*; poetry is almost totally lost in such a case. The point is rather that if Jaspers is right, myths are not interpretable into non-myth; 'Myths interpret each other.'

Bultmann lost that debate in practice. Those on Bultmann's side freely grant that as pastors, they continue to talk the New Testament language of Resurrection and a Second Coming: the mythical element is not eliminated but re-interpreted. The myth of bondage to and conviction of sin becomes the Existentialist myth of the stranger and nausea; the myth of redemption and second birth becomes the myth of freedom, encounter, and authenticity.

Parallel to this *de facto* consensus about the language of myth we find a rare philosophical consensus, between today's realists and empiricists, about the language of science. Early in this century there was indeed the idea that science can be 'demythologized' in some strict empiricist way. But that idea already had to be abandoned more than 50 years ago. The language of science cannot be reduced through 'operational definitions' or translation into a hygienic, pure observation language.

Philosophers are often slow to adapt to their own discoveries and advances, however. As a result, the sense of transition to a truly non-reductionist view of science did not become prevalent until the 1960s.[10] In the aftermath, the stamp of orthodoxy placed on this realization was somewhat confusingly associated with scientific realism, and equally with Feyerabend and Kuhn, who are also readable as critics of scientific realism. No distinction can be drawn between the theoretical and the non-theoretical, and there is not even in principle, however attenuated, any way to isolate a non-theoretical foundation for our conceptual framework. *Theoretical discourse is irreducible*. Theories can at most be interpreted in other, later theories; as Newton's mechanics was re-interpreted (as of restricted, approximate validity) by Einstein. Briefly: demythologizing the language of science is impossible.

This additional parallel between philosophical and theological debates throws a corollary light on a phenomenon we see both in science and in the varieties of religious experience: that of conceptual immersion. If the language to be used is not translatable without loss into something conceptually poorer, then to speak it we must allow ourselves to be guided by the entire picture presented. There is no disengaged alternative.

## 4 The scientific spirit

### 4.1 The demarcation of science and myth

After all this, you may think that at this point you know very well what my conclusion is going to be. You may very well think that now I am going to say:

(1) Meaningfulness always came from immersion/enchantment in a Great Myth – such as the Christian Myth in the Middle Ages in Europe;
(2) Science too is a Great Myth, providing us with a world-view able to replace the lost myths of the past;
(3) Meaning will be regained if we immerse ourselves now in this new Great Myth:
*Let the Scientific Middle Ages begin!*

Nothing is further from my mind. I do think that there is a great difference between science and the older myths – not an essential black-and-white difference, but still a difference in fact and of degree which is of enormous importance. And I think that scientific realists miss or obscure this difference exactly because they focus on *content* rather than *function*.

What is so crucially different about science? Let me quote again from Bultmann, who had some stake in this:

> The science of today is no longer the same as it was in the nineteenth century . . . The main point, however, is not the concrete results of scientific research and the contents of a world view, but the *method of thinking* from which world views follow.
>
> The contrast between the ancient world view of the Bible and the modern world view is the contrast between two ways of thinking, the mythological and the scientific. The method of scientific thinking [. . . is] the same in modern scientific research [but the theories] are changing over and over again, since the change itself results from the permanent principles [of this method].[11]

Karl Jaspers has already objected that there is a good deal of old and dated philosophy of science in Bultmann's writings. The caricature we can read into it is this: (1) in myth, content is what is important, and the commitment is to a world-picture; but in science, the method of inquiry is what matters and commitment is to a method; (2) within myth, questioning beyond a certain point is a sin; within science everything is subjected rigorously to proof and experimental test. This appealingly simple picture is almost entirely a mess of half-truths and propaganda. But I must emphasize the 'almost'. Let us count the ways in which all this is false.

(1) When we contrast science to myth and superstition, we are contrasting content, not method.

(2) No theory is *established* by proof, experimental or otherwise. (To accept a theory is *ipso facto* to go beyond the facts.)
(3) Scientists follow that scientific ethic in practice only to a limited extent (as Paul Feyerabend emphasizes in chapter 6). (Perhaps progress in science would also be impossible if the ethic of systematic doubt were not held in check in practice. How many geniuses can a science afford, in one century?)
(4) The attitude prescribed by the scientific ethic is also possible – perhaps to the same limited extent as in science – within myth in general. (Indeed, the major and radical changes to which the great myths too are subject argue that there is an underlying commitment which transcends, and is not indissolubly linked to, belief in particular content.)

However all this may be, I think Bultmann put his finger on the crucial aspect of science, in which in practice and in ethic it sets itself apart from all its actual rivals. *We should not exaggerate its extent, but we cannot exaggerate its importance.*

For this is the creed and regulative ideal of science: that our first and overriding commitment shall be *to the method*, uncompromisingly rigorous, which sweeps before it like chaff the inadequate structures of earlier hypotheses. The holy war in which the religious devotee systematically destroys the 'old man' (to use St Paul's term), uprooting one by one the binding desires and illusions in the soul, is transposed by science to its own growth. The primary commitment is to this method with its ideal of constant revaluation and self-critique; all commitment to content is secondary. This is the peculiarity of science among myths.

The hierarchy of responsibility is inverted; in the old myths, to avoid doubt may be piety, in the new it is treason.

### 4.2 Realism's mistaken moral

Scientific attitude as transvaluation of all values? Yes, that is what I mean.

If this conclusion about the primacy of method *vis à vis* content in science is correct, then realism has throughout mis-focussed the debate. For if realist metaphysicians reify content, then they do for science what the superstitious do for religion: *they avert attention from its significance to the vehicle of that significance.*

What this means is that acceptance of science, and appreciation of its worth, does not require us to believe that it is true. On the contrary, the important point about scientific activity is not that it provides theories which every generation in turn can take as truth, but rather that it accustoms us to giving up our beliefs, changing and altering them, valueing them without being in bondage to them.

Can we feel secure and at home in a world without the certainty that we know what it is really like?

This sounds like a psychological question; and in that form a philosopher has no business or interest in answering it. But actually I think that in a psychological sense the question does not arise at all: we have never had any objective certainty in our interpretation of what is 'really going on', and never will. We have seen the content of the scientific world-picture change radically, and we fully expect there to be more scientific revolutions in our future. So if a psychologist came along and told us we could not cope unless we have full belief in some specific world-picture, he'd just be advising refuge in self-deception.

The scientific attitude, which the empiricist celebrates, does not lead to despair and futility or disorientation. There is a loss when you first lose your certainties, and a temptation to seek refuge in some artificial certainty. When the refuge sought is in some part of science adopted as dogma, we call that *scientism*. That is, a sort of science on a metaphysical pedestal, with the current content of science erected into a final measure of all truth and value. Believe me, that is not science – it is superstition, no matter how scientific it is made to sound. To the extent that scientific realism shades into scientism, it has the same pitfall: to require the sacrifice of the intellect, the desperation of '*Credo ut intelligam*'.

What is the alternative to reifying the content of science? The alternative is to accept the challenge of intellectual maturity: to let your faith be not a dogma but a search, not an answer but a question and a quest, and to immerse yourself in a new world-picture without allowing yourself to be swallowed up.

Science allows perfectly well the sceptical discipline that accepts the appearances alone as real, and all the rest as a unifying myth to light our path.

## Notes

1. See e.g. Ockham, *Ord. Prol.* q. 9 and *Summa Totius Logicae* III. 2, 5; cf. M. H. Carré, *Realists and Nominalists* (Oxford University Press, 1946).
2. See Ockham, *Ord. Prol.* q. 6, J; *Summa Totius Logicae* III, II, 5, 64v, discussed by E. A. Moody, *The Logic of William of Ockham* (London, Russell & Russell, 1935, 2nd edn 1965). See further E. A. Moody, *Studies in Medieval Philosophy, Science, and Logic* (Berkeley, University of California Press, 1975).
3. Ockham, *Summulae in Libros Physicorum* III, 7. See also *Reportatio* II, q. xxvi N ff. which is printed together with an English translation in P. Boehner, *Ockham: Philosophical Writings* (London, Thomas Nelson, 1957), pp. 139–41, and discussions of Ockham's natural philosophy in Moody, *Studies* and E. J. Dijksterhuis, *The Mechanization of the World Picture* (Oxford, Clarendon, 1961).
4. Robert Silverberg, *The Stochastic Man* (London, Victor Gollancz, 1976).
5. Jean-Paul Sartre, *Nausea* (New York, *New Directions*, 1964), p. 106.
6. Ibid. pp. 211, 212.
7. See Ainslie Roberts and Ch. P. Mountford, *The First Sunrise: Australian aboriginal myths in paintings by Ainslie Roberts, with text by Ch. P. Mountford* (Adelaide, Rigby, 1971), pp. 9, 11. The omitted part between these quotations includes:
   'These myths also describe how, before creation times, the uncreated and eternal earth had always existed as a large flat disc floating in space . . . It was a dead, silent world. Yet, slumbering beneath that monotonous surface, were indeterminate forms of life that would eventually transform the forbidding landscape into the world as the aborigines know it today. As the ages passed, these mythical beings began to emerge from beneath the plain and to wander haphazardly over its surface.'
8. Rudolf Bultmann, 'The Testament and mythology' in H. W. Bartsch, *Kerygma and Myth* (London, SPCK, 2nd edn, in two vols, 1964), vol. 1 pp. 1–44 esp. 1,3.
9. Karl Jaspers, 'Truth and misfortune in Bultmann's mythologisation' in H. W. Bartsch, *Kerygma and Myth*, vol. 2, p. 144.
10. See for example the papers by Rudolf Carnap and Grover Maxwell in the first and third volumes of *Minnesota Studies in the Philosophy of Science*. Among the earlier papers leading up to this are Wilfrid Sellars, 'Concepts as involving laws and inconceivable without them' in *Philsophy of Science* 50 (1948), 287–315 and Paul Feyerabend, 'An attempt at a realistic interpretation of experience' in *Proceedings of the Aristotelian Society* NS 58 (1958), 143ff.
11. Rudolf Bultmann, *Jesus Christ and Mythology* (New York, Scribners, 1958), pp. 37, 38 (my italics).

# 6 Has the scientific view of the world a special status compared with other views?

Paul Feyerabend

Dennis Dieks, in Chapter 3, sketches a framework which, he says, has guided the work of many physicists. He implies that the remaining conflicts are a purely philosophical phenomenon. Being fond of quarrels philosophers have split into schools. There are now empiricists, positivists, rationalists, anarchists, realists, apriorists, pragmatists and they all have different views about the nature of science. Scientists, on the other hand, collaborate. Collaboration creates uniformity and, with it, a single way of looking at things: it *does* make sense to ask about the status of *the* scientific world-view.

In contrast I want to argue that scientists are as contentious as philosophers. But while philosophers merely talk, scientists act on their convictions; scientists from different areas use different procedures and construct their theories in different ways. Moreover, they often succeed: the world-views we find in the sciences have empirical substance. *This is a fact, not a philosophical position.* I shall explain it by considering the following four questions:

(1) What is *the* scientific view of the world and is there a single such view?
(2) Assuming there is a single scientific world-view – *for whom* is it supposed to be special?
(3) What kind of *status* are we talking about? Popularity? Practical advantages? Truth?
(4) What 'other views' are being considered?

My answer to the *first question* is that the wide divergence of individuals, schools, historical periods and entire sciences makes it difficult to identify comprehensive principles either of method, or of fact.

In the domain of *method* we have scientists like Luria who want to tie research to events permitting 'strong inferences', 'predictions that will be strongly supported and sharply rejected by a clear-cut experimental step'.[1]

According to Luria the experiments,[2] which showed that the resistance of bacteria to phage invasion is a result of environment-independent mutations and not of an adaptation to the environment, had precisely this character. There was a simple prediction; the prediction could be tested in a straightforward way yielding a decisive and unambiguous result. (The result refuted Lamarckism which was popular among bacteriologists but practically extinct elsewhere – a first indication of the complexity of science.)

Scientists of Luria's inclination show a considerable 'lack of enthusiasm in the 'big problems' of the Universe or of the early Earth or in the concentration of carbon dioxide in the upper atmosphere',[3] all subjects that are 'loaded with weak inferences'.[4] In a way they are continuing the Aristotelian approach which demands close contact with experience and objects to following a plausible idea to the bitter end.[5]

However this was precisely the procedure adopted by Einstein, by researchers in celestial mechanics between Newton and Poincaré (stability of the planetary system), by the proponents of atomism and, later, of the kinetic theory, by Heisenberg during the initial stages of matrix mechanics (when it seemed to clash with the tracks produced in the Wilson chamber) and by almost all cosmologists. Einstein's first cosmological paper is a purely theoretical exercise containing not a single astronomical constant. The subject of cosmology itself for a long time found few supporters among physicists. Hubble, the observer, was respected, the rest had a hard time:

> Journals accepted papers from observers, giving them only the most cursory refereeing whereas our own papers always had a stiff passage, to a point where one became quite worn out with explaining points of mathematics, physics, fact and logic to the obtuse minds who constitute the mysterious anonymous class of referees, doing their work, like owls, in the darkness of the night.[6]

'Is it not really strange,' asks Einstein[7] 'that human beings are normally deaf to the strongest argument while they are always inclined to overestimate measuring accuracies?' But just such an 'overestimating of measuring accuracies' is the rule in epidemiology, demography, genetics, spectroscopy and in other subjects. The variety increases when we move into sociology or cultural anthropology where a compromise has to be found between the effects of personal contact, the pressing needs of a region and the idea of objectivity. Robert Chambers, the inventor of the method of rapid rural appraisal writes as follows:

> To hear a seminar at the university about modes of production in the morning and then attend a meeting in a government office about agricultural extension in the afternoon leaves a schizoid feeling. One might not know that both referred to the same small farmers and might doubt whether either discussion has anything to contribute to the other.[8]

Methods are not restricted to particular areas. Luria's requirements seem to be tied to laboratory science and, more especially, to bacteriology; however they also turn up in astrophysics and cosmology. For example, they were used by Heber Curtis in 1921, in his 'grand debate' with Harlow Shapley. (Curtis doubted that stellar features could be as readily generalized as was assumed by Shapley – and he was right, especially in the decisive case of the Cepheides.) They guided the great Armenian astrophysicist Viktor Ambarzumjan and they are now being applied by Halton Arp, Margaret Geller and their collaborators. In Prandtl's lectures we read:

> The great growth in technical achievement which began in the nineteenth century left scientific knowledge far behind. The multitudinous problems of practice could not be answered by the hydrodynamics of Euler; they could not even be discussed. This was chiefly because, starting from Euler's equations of motion the science had become more and more a purely academic analysis of the hypothetical frictionless 'ideal fluid'. This theoretical development is associated with the names of Helmholtz, Kelvin, Lamb and Rayleigh.
>
> The analytical results obtained by means of this so called 'classical hydrodynamics' virtually do not agree at all with the practical phenomena . . . Therefore the engineers . . . put their trust in a mass of empirical data collectively known as the 'science of hydraulics', a branch of knowledge which grew more and more unlike hydrodynamics.[9]

According to Prandtl we have a disorderly collection of facts on the one side, sets of theories starting from simple but counterfactual assumptions on the other and no connection between the two. More recently the axiomatic approach in quantum mechanics and especially in quantum field theory was compared by clinical observers to the 'shakers',

> a religious sect of New England who built solid barns and led celibate lives, a non-scientific equivalent of proving rigorous theorems and calculating no cross sections.[10]

Yet in quantum mechanics this apparently useless activity has led to a more coherent and far more satisfactory codification of the facts than had been achieved before while in hydrodynamics 'physical commonsense' occasionally turned out to be less accurate than the results of rigorous proofs based on wildly unrealistic assumptions. An early example is Maxwell's calculation of the viscosity of gases. For Maxwell this was an exercise in theoretical mechanics, an extension of his work on the rings of Saturn. Neither he nor his contemporaries believed the result – that viscosity remains constant over a wide range of density – and there was contrary evidence. Yet more precise measurements turned the apparent failure into a striking success.[11]

Meanwhile the situation has changed in favor of theory. In the 1960s and 1970s, when science was still admired, theory got the upper hand, at universities where it increasingly replaced professional skills, and in special subjects such as biology or chemistry where earlier morphological and substance-related research was replaced by a study of molecules. In cosmology a firm belief in the Big Bang now tended to devalue observations that clashed with it. C. Burbidge writes:

> Such observations are delayed at the refereeing stage as long as possible with the hope that the author will give up. If this does not occur and they are published the second line of defence is to ignore them. If they give rise to some comment, the best approach is to argue simply that they are hopelessly wrong and then, if all else fails, an observer may be threatened with loss of telescope time until he changes his program.[12]

These and similar examples show that science contains different trends with different research philosophies. One trend requires that scientists stick closely to the facts, design experiments that clearly establish the one or the other of two conflicting alternatives and avoid far reaching speculations. One might call it the Aristotelian trend. Another trend encourages speculation and is ready to accept theories that are related to the facts in an indirect and highly complex way. Let us call this the Platonic trend. The existence of different trends within a comprehensive venture is not surprising. On the contrary, it would be strange if large groups of passionate and imaginative people, who despise authority and make criticism a guide to research, subscribed to a single point of view. What is surprising is that almost all the trends that developed within the sciences, Aristotelianism and an extreme Platonism included, produced results, not only in special domains, but everywhere; there exist highly theoretical branches of biology and highly empirical parts of astrophysics. The world is a complex and many-sided thing.

So far I have been talking about procedures, or methods. Now methods that are not used as a matter of habit, without any thought about the reasons behind them, are often tied to metaphysical beliefs. Aristotelians assume that humans are in harmony with the Universe; observation and truth are closely related. For Platonists humans are deceived in many ways. It needs abstract thought to get in touch with reality. Adding empirical success to these and other trends we arrive at the result that *science contains many different and yet empirically acceptable world-views, each one containing its own metaphysical background*. Johann Theodore Merz[13] describes the situation in the nineteenth century. He discusses the following views. *The astronomical view* rested on mathematical refinements of action-at-a-distance laws and was extended (by Coulomb, Neumann, Ampère and others) to electricity and magnetism. Laplace's theory of capillarity was an outstanding achievement of this approach. (In the eighteenth century Benjamin Rush had applied the view to medicine thus creating a unified but deadly medical system.) *The atomic view* played an important role in chemical research (e.g. stereochemistry) but was also opposed by chemists, for empirical as well as for methodological reasons. *The kinetic and mechanical view* employed atoms in the area of heat and electric phenomena. For some scientists (but by no means for all) atomism was the foundation of everything. *The physical view* tried to achieve universality in a different way, on the basis of general notions such as the notion of energy. It could be connected with the kinetic view, but often was not. Physicians, physiologists and chemists like Mayer, Helmholtz, du Bois-Reymond and, in the practical area, Liebig were outstanding representatives in the later nineteenth century while Ostwald, Mach and Duhem extended it into the twentieth. *The phenomenological view* (not mentioned by Merz) is related to the physical view, but less general. It was adopted by scientists like Lamé who found that it provided a more direct way to theory (elasticity, in the case of Lamé) than the use of atomic models. Starting his description of *the morphological view* Merz writes:

> The different aspects of nature which I have reviewed in the foregoing chapters and the various sciences which have been elaborated by their aid, comprise what may appropriately be termed the abstract study of natural objects and phenomena. Though all the methods of reasoning with which we have so far become acquainted originated primarily through observation and the reflection over things natural, they have this in common that they – for the purpose of examination – remove their objects out of the position and surroundings which nature has assigned to them: that they *abstract* them. This process of abstraction is either literally a process of removal from one place to another, from the great work – and storehouse of nature herself to the small work-room, the laboratory of the experimenter; or – where such removal is not possible – the process is carried out merely in the realm of contemplation; one or two special properties are noted and described, whilst the number of collateral data are for the moment disregarded. [A third method, not developed at the time is the creation of 'unnatural' conditions and, thereby, the production of 'unnatural' phenomena.]
>
> There is, moreover, in addition to the aspect of convenience, one very powerful inducement for scientific workers to persevere in their process of abstraction . . . This is the practical usefulness of such researches in the arts and industries . . . The wants and creations of artificial life have thus proved the greatest incentives to the abstract and artificial treatment of natural objects and processes for which the chemical and electrical laboratories with the calculating room of the mathematician on the one side and the workshop and factory on the other, have in the course of the century become so renowned . . .
>
> There is, however, in the human mind an opposite interest which fortunately counteracts to a considerable extent the one-sided working of the spirit of abstraction in science . . . This is the genuine love of nature, the consciousness that we lose all power if, to any great extent,

> we sever or weaken that connection which ties us to the world as it is – to things real and natural: it finds its expression in the ancient legend of the mighty giant who derived all his strength from his mother earth and collapsed if severed from her . . . In the study of natural objects we meet [therefore] with a class of students who are attracted by things as they are . . . [Their] sciences are the truly descriptive sciences, in opposition to the abstract ones.[14]

I have quoted this description at length for it shows very clearly how different interests leading to different procedures collect evidence which can then congeal into an empirically founded world-view. Finally, Merz mentions the genetic view, the psychophysical view, the vitalistic view, the statistical view together with their procedures and their findings.

What can a single comprehensive 'world-view of science' offer under such circumstances?

It can offer a survey, a *list* similar to the list given by Merz, enumerating the achievements and drawbacks of the various approaches as well as the clashes between them and it can identify science with this complex and somewhat scattered war on many fronts. Alternatively it can put one view on top and *subordinate* the others to it, either by pseudo-derivations, or by declaring them to be meaningless. Reductionists love to proceed in this fashion. Or it can disregard the differences and present a 'paste job' where each particular view and the results it has achieved is smoothly connected with the rest, thus producing an impressive and coherent edifice – *the* scientific world-view.

Expressing it differently we may say that the assumption of a single coherent world-view that underlies all of science is either a metaphysical hypothesis trying to anticipate a future unity, or a pedagogical fake; or it is an attempt to show, by a judicious up- and downgrading of disciplines, that a synthesis has already been achieved. This is how fans of uniformity proceeded in the past (cf. Plato's list of subjects in Chapter seven of his *Republic*), these are the ways that are still being used today. A more realistic account, however, would point out that

> [t]here is no simple 'scientific' map of reality – or if there were, it would be much too complicated and unwieldy to be grasped or used by anyone. But there are many different maps of reality, from a variety of scientific viewpoints.[15]

You may object that we live in the twentieth century, not in the nineteenth and that unifications which seemed impossible then have been achieved by now. Examples are statistical thermodynamics, molecular biology, quantum chemistry and superstrings. These are indeed flourishing subjects, but they have not produced the unity the phrase '*the* scientific view of the world' insinuates. In fact, the situation is not very different from what Merz had noticed in the nineteenth century. Truesdell and others continue the physical approach. Prandtl maligned Euler; Truesdell praises him for his rigorous procedure. Morphology, though given a low status by some and declared to be dead by others, has been revived by ecologists and by Lorenz' study of animal behavior (which added forms of motion to the older static forms); it always played an important role in galactic research (Hubble's classification) and astrophysics (Hertzsprung–Russell diagram). Having been in the doghouse cosmology is now being courted by high energy physicists but clashes with the philosophy of complementarity accepted by the same group. Commenting on the problem M. Kafatos and R. Nadeu write:

> The essential requirement of the Copenhagen interpretation that the experimental setup must be taken into account when making observations is seldom met in observations with cosmological import [though such observations rely on light, the paradigm case of complementarity].[16]

Moreover, the observations of Arp, M. Geller and others have thrown doubt on the homogeneity assumption which plays a central role in cosmology. We have a rabid materialism in some parts (molecular biology, for example), a modest to radical subjectivism in others (some versions of quantum measurement, anthropic principle). There are many fascinating results, speculations, attempts at interpretation and it is certainly worth knowing them. But pasting them together into a single coherent 'scientific' world-view, a procedure which has the blessings even of the Pope[17] – this is going too far. After all – who can say that the world which so strenuously resists unification really is as educators and metaphysicians want it to be – tidy, uniform, the same everywhere? Besides, a 'paste job' eliminates precisely those conflicts that kept science going in the past and will continue inspiring its practitioners if preserved.

I now come to the *second question*: special status – for whom? I do not want to argue that considerations of status (truth, reality) are necessarily relativistic. But social considerations do play a role and they have occasionally advanced the cause of science. Consider the following example (which involves far-reaching interpretations from slender data).

Like modern scientists the Ionian natural philosophers (Thales, Anaximander, Anaximenes) looked for simple principles behind the variety of phenomena. Today the principles sought are theories or laws. In ancient Greece they were substances. According to Thales water was the basis of everything. This is not as implausible as it sounds. The Greeks who 'lived around the Mediterranean like frogs around a pond' could *see* how water transformed, first, into mist, then into air and perhaps even into fire (lightning). Frozen water was solid (earth) and, besides, water was needed everywhere, to sustain life. Using a symmetry principle Anaximander objected that fire, earth and air seemed to be as important as water which means that the basic substance had to be different from all elements, though capable of turning into them under special circumstances. Anaximander called it *apeiron* – the unlimited. Parmenides then pointed out that Being was still more fundamental (water *is*, fire *is*, *apeiron is* – they are all forms of Being). What can be said about Being? That it is and that not-Being is not. Note that the statement BEING IS (*estin* in the Greek of Parmenides) was the first explicit conservation principle of the West: it asserted the conservation of Being. Accepting this argument we can infer that there is no change: the only possible change is into not-Being, not-Being does not exist, hence there is no change. What about difference? The only possible difference is between Being and not-Being, not-Being does not exist, hence Being is everywhere the same. But don't we perceive change and difference? Yes, we do, which shows that change and difference are appearances, chimeras. Reality does not change. This was the first and most radical (Western) theory of knowledge. It is not entirely ridiculous: nineteenth-century science up to and including Einstein also devalued change. Hermann Weyl writes:

> The relativistic world simply *is*, it does not *happen*. Only to the gaze of my consciousness, crawling upward along the lifeline of my body, does

> a section of this world come to life as a fleeting image in space which continuously changes in time.[18]

Ancient atomism can be seen as an attempt to shorten the distance between basic physics (BEING IS) and common sense. Leukippos and Democritos retained one part of Parmenides' theory (atoms are tiny fragments of Parmenidean Being) and rejected another (not-Being exists and it is identical with space). There was no refutation – after all, Parmenides had proved that change was apparent and you cannot refute a theory about real things by confronting it with appearances. The aim was, rather, to adapt physics to certain social tendencies (such as the tendency to regard change as something very important). In Aristotle this aim is stated explicitly: real is what plays a basic role in the life we want to lead. The second question now amounts to this: are we prepared to view ourselves in the manner suggested by scientists or do we prefer to make personal contact, friendship etc. the measure of our nature? Note that what is needed here is a personal (social) decision, not a scientific argument. Instrumentalists explain how we can have our cake, i.e. how we can continue holding familiar views (religious views, for example), and eat it too (i.e. profit from the practical results of the sciences).

*Third question*: what do we mean by status? *Popularity*, i.e. familiarity with major results and the admission that they are important, would be one measure. Now it is true that despite periodic swings towards the sciences and away from them the sciences are still in high repute with the general public. Or, to be more precise, not the sciences are, but a mythical monster 'science' (in the singular – in German it sounds even more impressive DIE WISSENSCHAFT). For what the so-called 'educated public' seems to assume is that the achievements they read about in the educational pages of their newspapers and the threats they seem to perceive come from a single source and are produced by a uniform procedure. I have tried to argue that scientific practice is much more diverse. Adding that 10% of all average Germans still believe in a stable Earth, that a third of all adults in the USA believe everything in the Bible to be literally true,[19] that a promotional page for a new journal, *Public Understanding of Science* contains the passage:

> Annual polls show again and again a yawning gap between public attitudes and scientific advances. In a Gallup poll commissioned this summer [1991] a picture emerges of a population which not only confesses ignorance but also a substantial lack of concern about the giant discoveries transforming everyday life.

and that the 'population' referred to is the Western middle class, not Bolivian peasants (for example) I conclude that popularity can hardly be regarded as a measure of excellence.

What about *practical advantages?* The answer is that 'science' sometimes works and sometimes doesn't but that it is still sufficiently mobile to turn disaster into triumph. *It can do that because it is not tied to particular methods or world-views.* The fact that an approach or a subject (economics, for example) is 'scientific' according to some abstract criterion is therefore no guarantee that it will succeed. *Each case must be judged separately,* especially today, when the inflation of the sciences has added some rather doubtful activities to what used to be a sober enterprise.[20] The question of *truth,* finally, remains unresolved. Love of Truth is one of the strongest motives for replacing what really happens by a streamlined account or, to express it in a less polite manner – love of truth is one of the strongest motives for deceiving oneself and others. Besides, the quantum theory seems to show, in the precise manner beloved by the admirers of science, that reality is either one, which means there are no observers and no things observed, or it is many, including theoreticians, experimenters and the things they find, in which case what is found does not exist in itself but depends on the procedure chosen.

I now come to the *last question*: what are the other views that are being considered? In my public talk I quoted a passage from E. O. Wilson.[21] It runs as follows:

> . . . religion . . . will endure for a long time as a vital force in society. Like the mythical giant Antaeus who drew energy from his mother, the earth, religion cannot be defeated by those who may cast it down. The spiritual weakness of scientific naturalism is due to the fact that it has no such primal source of power . . . So the time has come to ask: does a way exist to divert the power of religion into the services of the great new enterprise?

For Wilson the main feature of the alternatives is that they have *power*. I regard this as a somewhat narrow characterization. World-views also answer questions about origins and purposes which sooner or later arise in almost every human being. Answers to these questions were available to Kepler and Newton and were used by them in their research; they are no longer available today, at least not within the sciences. They are part of non-scientific world-views which therefore have much to offer, also to scientists. When Western civilization invaded the Near and Far East and what is now called the Third World it imposed its own ideas of a proper environment and a rewarding life. Doing this, it disrupted delicate patterns of adaptation and created problems that had not existed before. Human decency and an appreciation of the many ways in which humans can live with nature have prompted agents of development and public health to think in more complex or, as some would say, more 'relativistic' ways – which is in complete agreement with the pluralism of science itself.

To sum up: there is no 'scientific world-view' just as there is no uniform enterprise 'science' – except in the minds of metaphysicians, schoolmasters and scientists blinded by the achievements of their own particular niche. Still, there are many things we can learn from the sciences. But we can also learn from the humanities, from religion and from the remnants of ancient traditions that survived the onslaught of Western civilization. No area is unified and perfect, few areas are repulsive and completely without merit. There is no objective principle that could direct us away from the supermarket 'religion' or the supermarket 'art' towards the more modern, and much more expensive supermarket 'science'. Besides, the search for such guidance would be in conflict with the idea of individual responsibility which allegedly is an important ingredient of a 'rational' or scientific age. It shows fear, indecision, a yearning for authority and a disregard for the new opportunities that now exist: we can build world-views on the basis of a personal choice and thus unite, for ourselves and for our friends, what was separated by the chauvinism of special groups.

On the other hand, we can agree that in a world full of scientific products scientists may be given a special status just as henchmen and generals had a special status at times of social disorder or priests when being a citizen coincided with being the member of a church. We can also agree that appealing to a chimera can have important political consequences. In 1854 Commander Perry, using force, opened the ports of Hakodate and Shimoda to American ships for supply and trade. This event demonstrated the military inferiority of Japan. The members of the Japanese enlightenment of the early 1870s, Fukuzawa among them, now reasoned as follows: Japan can keep its independence only if it becomes stronger. It can become stronger only with the help of science. It will use science effectively only if it does not just practice science but also believes in the underlying ideology. To many traditional Japanese this ideology – 'the' scientific world-view – was barbaric. But, so the followers of Fukuzawa argued, it was necessary to adopt barbaric ways, to regard them as advanced, to introduce the whole of Western civilization in order to survive.[22] Having been thus prepared, Japanese scientists soon branched out as their Western colleagues had done before and falsified the uniform ideology that had started the development. The lesson I draw from this sequence of events is that a uniform 'scientific view of the world' may be useful *for people doing science* – it gives them motivation without tying them down. It is like a flag. Though presenting a single pattern it makes people do many different things. However, *it is a disaster for outsiders* (philosophers, fly-by-night mystics, prophets of a New Age, the 'educated public'), who, being undisturbed by the complexities of research, are liable to fall for the most simple-minded and most vapid tale.

## Notes

1. S. E. Luria, *A Slot Machine, a Broken Test Tube* (Harper & Row, New York, 1984), p. 115.
2. S. E. Luria and M. Delbrück, in *Genetics* **28** (1943), 491.
3. Luria, *A Slot Machine*, p. 119.
4. *Ibid.*
5. Aristotle, *In De Caelo*, 293 a 25ff.
6. F. Hoyle, in Y. Terzian and S. M. Bilson (eds.), *Cosmology and Astrophysics* (Cornell University Press, London, 1982), p. 21.
7. *The Born–Einstein Letters* (Walter, New York, 1971), p. 192.
8. Robert Chambers, *Rural Development* (Longman, London, 1986), p. 29.
9. L. Prandtl and O. G. Tietjens, *Fundamentals of Hydro- and Aerodynamics* (Dover, New York, 1957), p. 3.

10. R. F. Streater and A. S. Wightman, *PCT, Spin, Statistics and all that* (W. A. Benjamin, New York, 1964), p. 1.
11. For quantum mechanics see Sections 4.1 and 4.2 of Hans Primas, *Chemistry, Quantum Mechanics and Reductionism* (Springer, Berlin, 1981). Maxwell's calculations are reproduced in *The Scientific Papers of James Clerk Maxwell*, W. D. Niven (ed.) (Dover, New York, 1965, first published in 1890), pp. 377ff. The conclusion is stated on p. 391: 'A remarkable result here presented to us . . . is that if this explanation of gaseous friction be true, the coefficient of friction is independent of the density. Such a consequense of mathematical theory is very startling, and the only experiment I have met with on the subject does not seem to confirm it.' For examples from hydrodynamics see G. Birkhoff, *Hydrodynamics* (Dover, New York, 1955), Sections 20 and 21.
12. C. Burbidge, 'Problems of cosmogony and cosmology' in F. Bertola, J. W. Sulentic, D. F. Madore (eds.), *New Ideas in Astronomy* (Cambridge University Press, 1988), p. 229.
13. Johann Theodore Merz, *A History of European Thought in the 19th Century* (Dover, New York, 1965, first published 1904–12).
14. Merz, *A History*, Vol. 2, p. 200f.
15. John Ziman, *Teaching and Learning about Science and Society* (Cambridge University Press, 1980), p. 19.
16. M. Kafatos and R. Nadeu, 'Complementarity and cosmology' in M. Kafatos (ed.), *Bell's Theorem, Quantum Theory and Conceptions of the Universe* (Kluwer Academic Publishers, 1980), p. 263, see also Wheeler's thought experiment involving a quasar situated behind a galaxy.
17. See Pope John Paul II's message on the occasion of the 300th anniversary of Newton's *Principia*, published in *John Paul II on Science and Religion*, ed. R. J. Russell, W. R. Stoeger SJ and G. V. Coyne (Vatican Observatory Publications, 1990), especially M6ff.
18. Hermann Weyl, *Philosophy of Mathematics and Natural Science* (Princeton University Press, 1949), p. 116.
19. *International Herald Tribune* (December 24/5 1991), p. 4.
20. This was realized by government advisers after the post-war euphoria had worn off. 'The idea of a comprehensive science policy was gradually abandoned. It was realized that science was not one but many enterprises and that there could be no single policy for the support of all of them.' (Joseph ben-David, *Scientific Growth* (University of California Press, 1991), p. 525.)
21. E. O. Wilson, *On Human Nature*, (Havard University Press, 1978).
22. Details in Carmen Blacker, *The Japanese Enlightenment* (Cambridge University Press, 1969). For the political background see Chapters 3 and 4 of Richard Storry, *A History of Modern Japan* (Penguin, Harmondsworth, 1982).

# 7 Quantum theory and our view of the world

Paul Feyerabend

When I was a student in Vienna, in the late 1940s, we had three physicists who were known to a wider public: Karl Przibram, Felix Ehrenhaft and Hans Thirring. Przibram was an experimentalist, a pupil of J. J. Thomson whom he often mentioned with reverence. Philosophers of science know him as the editor of a correspondence on wave mechanics between Schrödinger, Lorentz, Planck and Einstein. He was the brother of Hans Przibram, the biologist, and, I believe, the uncle of the neurophysiologist Karl Przibram. He talked with a subdued voice and wrote tiny equations on the blackboard. Occasionally his lectures were interrupted by shouting, laughing and trampling from below; that was Ehrenhaft's audience.

Ehrenhaft had been professor of theoretical and experimental physics in Vienna. He left when the Nazis came; he returned in 1947. By that time many physicists regarded him as a charlatan. He had produced and kept producing evidence for subelectrons, magnetic monopoles of mesoscopic size and magnetolysis, and he held that the inertial path was a spiral, not a geodesic. His attitude towards theory was identical with that of Lenard and Stark whom he often mentioned with approval. He challenged us to criticize him and laughed when he realized how strongly we believed in the excellence of say, Maxwell's equations without having calculated and tested specific effects. During a summer school in Alpbach he set up his experiments in a little farmhouse and invited everyone to have a look. Leon Rosenfeld was there; so was Maurice Pryce, one of the most abrasive physicists of his generation. They went in; when they reappeared they looked as if they had seen something obscene. However all they could say was 'obviously a *Dreckeffekt*'. Afterwards, in Ehrenhaft's lecture, Rosenfeld and Pryce sat in the front row. Having described his experiments Ehrenhaft went up to them and exclaimed: 'Was können sie sagen mit allen ihren schönen Theorien? Nichts können sie sagen. Still müssen sie sein. Sitzen müssen sie bleiben.' (What can they say with all their fine theories? They can say nothing. They must be silent. They must remain seated.) And, indeed, Rosenfeld and Pryce, so eloquent on other occasions, did not say a single word. Ehrenhaft may not have been mainstream. But he made us think – more than many mainstream scientists before and after him.

The third physicist was the theoretician Hans Thirring, the father of Walter Thirring and discoverer of the (general-relativistic) effect of a rotating shell on objects inside. Thirring was a pacifist and a friend of Einstein, Freud and other disreputable people. He was sacked when Austria became part of the Reich. When he returned he had completed a huge manuscript on the psychological foundations of world peace. 'This is important', he said; 'physics is not'. Later, as a member of parliament he suggested that Austria abandon its army and have its borders guaranteed by the neighboring powers; needless to say this suggestion did not get anywhere. Thirring had strong convictions; yet he never lost his sense of humor. For him the events in Germany and the war afterwards were signs of human folly, not of evil incarnate. He was a rare combination, a committed humanitarian *and* a sceptic.

Przibram, Ehrenhaft and Thirring differed in many ways. They were united in their belief that world-views are dangerous things and that physics should do without them. The belief had been held before – by Mach, Boltzmann, Franz Exner and his group and by the members of the Vienna Circle (a group of logical positivists centred on Vienna University in the 1920s and 1930s). Only a few of these thinkers realized that they were guided by a rival world-view which, though humanitarian in intention, disregarded important features of human existence.

Today the situation is much less uniform. Many scientists now raise the question of religion; others are looking for ways of making science as powerful as religion used to be. I shall try to clarify this development by considering to what extent world-views affected and perhaps even created scientific knowledge.

## 1 The nature of world-views

In his essay *On Human Nature*, E. O. Wilson, whom some regard as the father of sociobiology, writes about religion as follows.

> Religion will endure for a long time as a vital force in society. Like the mythical giant Antaeus who drew energy from his mother, the earth, religion cannot be defeated by those who may cast it down. The spiritual weakness of scientific naturalism is due to the fact that it has no such primal source of power . . . So the time has come to ask: does a way exist to divert the power of religion into the services of the great new enterprise?[1]

According to Wilson there is religion, and there is this 'great new enterprise', scientific naturalism. Both look at the world in a rather general way. But while scientific naturalism provides information and makes practical suggestions, religion in addition is a 'primal [i.e. a non political] source of power'.

Jacques Monod says the same, though with greater understanding of the problems involved.

> Cold and austere, proposing no explanation but imposing an ascetic renunciation of all other spiritual fare [the idea that objective knowledge is the only authentic source of truth] was not of a kind to allay anxiety but aggravated it instead. By a single stroke it claimed to sweep away the tradition of a hundred thousand years, which had become one with human nature itself. It wrote an end to the ancient animist covenant between man and nature, leaving nothing in place of that precious bond but an anxious quest in a frozen universe of solitude. With nothing to recommend it but a certain puritan arrogance, how could such an idea win acceptance? It did not; it still has not. It has however commanded recognition; but that it did solely because of its prodigious power of performance.[2]

According to Monod objectivism – and by this Monod means knowledge not involving aims and purposes – has great achievements to its credit. However it provides only part of what is needed for a full life. Asserting that there is nothing more to be had it produced a meaningless world. Or to repeat a much commented-upon statement by Steven Weinberg – the more the universe seems comprehensible, the more it seems pointless.

My third quotation is from a letter Wolfgang Pauli wrote to his colleague Markus Fierz. I shall say more about Wolfgang Pauli below. Here I only present the quotation.

> Science wants to examine reality. This problem is closely connected with the other problem, namely, *the idea of reality*. When speaking about reality the layman usually assumes that he is talking about something that is obvious and known. However to me it is an important and very difficult task of our times to work at building a new idea of reality. This is also what I mean when emphasizing . . . that science and religion must in some way be related to each other.[3]

Pauli is looking for a point of view that pays attention to science but also goes beyond it.

Taking these three quotations as my guide I shall define a *world-view* as a collection of beliefs, attitudes and assumptions that involves the whole person, not only the intellect, has some kind of coherence and universality and imposes itself with a power far greater than the power of facts and fact-related theories.

## 2 The power of world-views

Being constituted in this manner world-views have tremendous strength. They prevail despite the most obvious contrary evidence and they increase in vigor when meeting obstacles.[4] Cruel wars, deadly epidemics that killed people indiscriminately, natural catastrophes, floods, earthquakes, widespread famines could not overcome the belief in an all-powerful, just and even benign creator god. Altogether it seems that people who are guided by world-views are incapable of learning from experience.

For enlightened people this apparent irrationality is one of the strongest arguments against all forms of religion. What they fail to realize is that *the rise of the sciences depended on a blindness, or obstinacy of exactly the same kind.* Surrounded by comets, new stars, plagues, strange geological shapes, unknown illnesses, irrational wars, biological malformations, meteors, oddities of weather, the leaders of Western science asserted the universal, 'inexorable and immutable' character of the basic laws of Nature.[5] Early Chinese thinkers had taken the empirical variety at face value. They had favored diversification and had collected anomalies instead of trying to explain them away.[6] Aristotelians had emphasized the local character of regularities and insisted on a classification by multiple substances and corresponding accidents. Natural is what happens always or almost always, said Aristotle.[7] Scientists of an equally empirical bent, Tycho Brahe among them, regarded some anomalous events of their time as miracles or, as we might say today, they took cosmic idiosyncrasies seriously; others, like Kepler, ascribed them to the subjective reactions of the telluric soul while the great Newton, for empirical as well as for theological reasons, saw in them the finger of God.[8]

Even instrumentalism, the doctrine, that is, that scientific theories do not describe reality but are instruments for the prediction of what can be observed, had a metaphysical or, as one might say, a world-view backing. According to St Thomas (*Summa Theol.*, Question 8, article 4) universals existed side by side with God, only in a different manner,[9] while Duns Scotus and William of Ockham objected that all we have are the results of God's acts of will.[10] We do not understand why these acts occurred, we do not know what acts will occur in the future, we can only observe their results, connect them and hope for the best. It needed a tremendous act of faith or, to express it in the terminology I am using here, it needed guidance by a powerful world-view not only to *assume* regularities where no regularities had yet been found, but *to work towards them* for centuries, and in the face of numerous failures.

The result that emerged in the nineteenth century[11] was not a coherent science but a collection of heterogeneous subjects (optics, acoustics, hydrodynamics, elasticity, electricity, heat etc. in physics; physiology, anatomy etc. in medicine; morphology, evolutionism etc. in biology – and so on). Some of the subjects (hydrodynamics, for example) had only a tenuous relation to experiment, others were crudely empirical. It did not matter. Being firmly convinced that the world was uniform and subjected to 'inexorable and immutable laws' leading scientists interpreted the collection as an *appearance* concealing a uniform material *reality*.

With the notion of reality I come to the main topic of this chapter which is the relation of human achievements to a world whose features are independent of thought and perception or, to express it more dramatically, the idea that humans are aliens, not natural inhabitants of the universe.

## 3 Humans as aliens in a strange world

The idea assumes that the world is divided into (at least) two regions – a primary region consisting of important events and a secondary region that differs from the primary region, blocks our vision, is deceptive and, in many cases, evil. Grand dichotomies of this kind are found in many though by no means in all cultures. The dichotomies can be sharpened and made absolute by accidents, social tendencies, abstract reasoning and other agencies. The Gnostic movement, for example, occurred at a time of uncertainty when humans seemed subjected to irrational political and cosmic forces and when help seemed far away. Here are humans 'as they really are', i.e. their souls are imprisoned in bodies and the bodies in turn are imprisoned in a material cosmos. This double imprisonment, effected by low level demons, prevents humans from discovering the truth: the more information they possess about the material world, the more they get involved in it, the less they know. Revelation frees them from their predicament and gives them genuine knowledge.[12]

Realism is part of Genesis. *First* God created the material universe, *then* he created man and woman. The Fall erected a barrier between humans and the world: having been in harmony with nature they are now separated from it. The debates about the nature of Christ (Arianism, Docetism etc.) deal with the size of the gap.

Greek philosophy is different in some respects, similar in others. Traditional Greek religion distinguished between human and divine properties but without making the distinction absolute: the Greek gods have human features and they participate in human affairs. The distinction hardened when divinities merged as a result of travel and cultural exchange. The gods now lost in individuality but gained in power. Taking his lead from this development Xenophanes introduced a God-Monster that does not walk around, does not speak, feels no sympathy but, like a true intellectual, moves the universe by thought alone.[13] Parmenides eliminated the last human property – his principle is pure Being. Using rules familiar from the practice of Near Eastern law he deduced that Being is, that not-Being is not and that change and difference are illusions. Note that the statement 'Being *is*' (*estin* in Parmenides' Greek) was the first conservation law of the West: it asserted the conservation of Being. Like the Gnostics, Parmenides based his results on revelation: it is a goddess that tells him how to proceed.

Now it is interesting to see to what extent the division between two fundamentally different cosmic regions invaded the sciences. Nineteenth-century classical physics – and by this I mean what was then regarded as the fundamental science, classical point mechanics – posited a 'real' world without colors, smells, sounds and a minimum of change; all that happens is that certain configurations move reversibly from one moment to another. In a relativistic world even these events are laid out in advance. Here the world

> simply *is*, it does not *happen*. Only to the gaze of my consciousness, crawling upward along the lifeline of my body, does a section of this world come to life as a fleeting image in space which continuously changes in time.[14]

Or, to quote Einstein:

> For us who are convinced physicists, the distinction between past, present, future has no other meaning than that of an illusion, though a tenacious one.[15]

Note how Einstein's realistic world-view interferes with his empiricism. According to Einstein theories must be tested by experiment. Experiments are temporal processes and, therefore, illusions. But how can illusions inform us about reality? We see here to what extent some modern scientists repeat age-old and, to many, rather disreputable patterns of thought. But remember: modern science with its 'inexorable and immutable' laws could not have arisen without these patterns. This is how world-views can both further and hinder the sciences, depending on circumstances.

## 4 Realism as a world-view and as a scientific hypothesis

There is a widespread rumor that realism – the idea that the world as laid out in space and time is independent of human perception, thought and action – has been refuted by delicate but conceptually robust experiments.

Now if what I have said about world-views (remember my definition at the end of Section 1!) is correct, then the 'realism' of the rumor cannot possibly be a world-view. There is no fact, no series of facts, no highly confirmed theory that can dislocate the assumption, made by Einstein, that the events of our lives, experiments included, are nothing but illusions. And even this statement is not adequate. Being tied to individuals and groups a world-view cannot be 'Platonized' – it cannot be presented as a person-independent entity that enters into relations with other person-independent entities such as facts and/or theories; it has to be related to the individuals and the communities that are affected by it. And a community holding realism as a world-view simply cannot be shaken by contrary *facts*. If it *is* shaken then this means that it is already breaking up or that the facts presented are part of a powerful rival *world-view*.

Let me illustrate the situation with a historical example. Parmenides' conclusion – there is no change – was contested by his philosophical successors. The situation is usually presented in the following way: Parmenides, positing Being as a single principle inferred that there is no change. There is change, hence there has to be more than one principle. Democritos assumed infinitely many principles, Aristotle assumed two, potentiality and actuality. The story makes it appear that Parmenides, involved in deep speculation, overlooked change while Democritos, more a man of the world, paid attention to it. But Parmenides was not only familiar with change; in the second part of his poem he even explained some of its features. However he added that change was not real. To refute him one had to do more than point to changing things. One had to show that change was at least as basic as Being or, to express it differently, one had to show that those addressed by Parmenides possessed a world-view where change played a fundamental role.

This was exactly what Aristotle did. All of us, he said, are citizens – we participate in the actions and deliberations of the city-state; some of us are physicians who try to heal people, others are cobblers, they make boots – our existence is shot through with change and it is this existence that counts, not the results of speculation. Now – can we say that the recent experimental results, which allegedly refute the idea of an objective Nature laid out in space and time, affect scientists and the general public as strongly today as Greek common sense affected Athenians at the time of Aristotle? I don't think so: most scientists are practical realists. At any rate – the matter must be examined; it cannot be taken for granted.

On the other hand we can interpret realism as a hypothesis and not as a world-view. There are passages where Einstein formulates realism in this more modest and more 'scientific' manner.

> One cannot simply ask, 'Does a definite moment for the decay of a simple atom exist?' but rather instead, 'Within the framework of the total theoretical construction, is it reasonable to assume the existence of a definite moment for the decay of a simple atom?'[16]

In other words – can the idea of such a definite moment be incorporated into a theoretical framework that agrees with tests and basic principles? In the passage I have just quoted Einstein does not specify the nature of the tests required. He simply refers to scientific practice. He even rejects, and with some vehemence,[17] the idea that it is possible to compare a theory with a theory-independent 'reality': science, not metaphysical speculation is to decide the question.

In the opinion of most scientists science *has* decided the question – against Einstein's hypothesis. It is somewhat ironical that the decision was prepared by three passionate realists: by Einstein himself who in his argument of 1935 used an example especially suited for demonstrating the holistic character of quantum mechanics, by Schrödinger who first noticed this property of the example and by Bell whose theorem considerably simplified the demonstration.

Many scientists for whom realism was a world-view and not just a hypothesis now found themselves in a somewhat paradoxical situation. On the one hand they wanted to remain scientists which means they were inclined to read realism as a hypothesis. On the other hand they had a strong realistic bias. Now science does allow for the continued pursuit of apparently refuted ideas. The survival and eventual triumph of atomism, of the idea that the Earth moves, of conservation laws or, more recently, of matrix mechanics (in the face of well-defined particle tracks) depended on this possibility. But for world-view realists mere possibilities are not good enough. They want stronger support. Trying to obtain it they use phrases that hint at world-views without bringing them out of the closet. In what may have been his last interview, John Bell said:

> For me, it is so reasonable to assume that the photons in those experiments carry with them programs that have been correlated in advance, telling them how to behave. This is so rational that I think, that when Einstein saw that and the others refused to see it, *he* was the rational man. The other people, though history has justified them, were burying their heads in the sand. I feel that Einstein's intellectual superiority over Bohr, in this instance, was enormous, a vast gulf between the man who saw clearly what was needed, and the obscurantist. So, for me it is a pity that Einstein's idea does not work. The reasonable thing just does not work.[18]

'It is so reasonable to assume,' says Bell; Einstein 'was the rational man', Bohr 'the obscurantist'. What does that mean? It means that there is a world-view somewhere; that, presented in its full splendor, it would not sound very scientific – and that it is therefore hidden. It is hidden, yes – but from its hiding place it still affects the debate through insinuations, slogans and attitudes. Most scientific arguments about realism have this truncated character. To complete them we have to reveal the underlying world-view and its relation to the realistic hypothesis. In addition we have to embed the troubling experiments into a rival world-view that is stronger than special professional subjects, gives us a reason to rely on them and agrees with or even demands the negative outcome of the experiments. Niels Bohr's idea of complementarity contains a sketch of a non-realistic world-view that satisfies these requirements. Wolfgang Pauli tried to give a more detailed and more complete account.

## 5 Wolfgang Pauli's attempt to combine science and salvation[19]

Wolfgang Pauli was one of the leading physicists of his time. He won the Nobel Prize (for his discovery of the exclusion principle), wrote two surveys which have remained classics (one on relativity, the other on wave mechanics) and technical articles. Substantial suggestions and criticisms dealing with almost all aspects of the physics of his time are found in his letters, even on postcards. His correspondence had an enormous influence.[20] Some colleagues called him the conscience of physics: others, who seem to have suffered more, its scourge. Pauli delighted in pure knowledge; he despised applications – the 'dark backside of science' as he called them – and was highly critical of 'the merely rational from whose background a will to power is never entirely absent'. He had strong views about world events but refused to participate in collective enterprises.

> My impact should lie in how I *live*, what I *believe* and in the ideas I communicate in a more or less direct way to a small circle of pupils and acquaintances – and not in speaking out before a large public.[21]

Bohr favored political interventions of scientists. Pauli replied:

> Whoever wants to oppose the 'Will to Power' in a more spiritual way must not succumb to this Will to such an extent that he ascribes to himself a greater influence on world history than he actually has.

Or, in another letter to Bohr (Pauli's English);

> My attitude is therefore, that we have to be satisfied with the fact – well established by history – that ideas always had a great influence on the course of history and also on politicians, but it is better if we leave the direct actions in politics to other persons and remain on the periphery and not in the center of this dangerous and disagreeable machinery. In [this] attitude . . . I am – last not least – also influenced by the philosophy of Laotse, in which so much emphasis is laid on the indirect action, that his ideal of a good ruler is one whom one does not consciously notice at all.

Pauli wanted to spread his ideas by personal contact with well prepared friends, not by starting a movement.

This precise individualist was deeply concerned about the direction science had taken since the seventeenth century. To orient himself he examined traditions professional rationalists dismiss with a contemptuous shrug. He made two points; first, that the rise of modern science was based on a new cosmic feeling and not on experience alone.

Kepler, for example, started from the Trinity and ended up with natural laws:

> because he looks at the sun and the planets with the archetypal image in the background he believes with religious fervor in the heliocentric system – *by no means the other way around.*[22]

Pauli's second point was that the development had two branches; one separated humans from a world they were trying to understand and to control, the other sought salvation through a practice (astrology, alchemy, hermeticism) that involved matter and spirit on equal terms.[23] According to Pauli the second branch soon fell apart.

> Today [1955] rationalism has passed its peak. One definitely feels that it is too narrow . . . It seems plausible to abandon the merely rational in whose background the Will to Power always played an important role and to adopt its opposite, for example a Christian or Buddhist mysticism. However I believe that for those who are no longer convinced by a narrow rationalism and whom the spell of a mystical outlook, an outlook that views the external world with its oppressive multitude of events as a mere illusion fails to affect with sufficient strength, I feel that for such people there remains only one choice: to expose themselves fully to these . . . opposites and to their conflict. This is how a scientist can find an inner path to salvation.[24]

Finding a world-view, Pauli seems to say, is a personal matter that must be fought through by every individual; it cannot be settled by 'objective' arguments. But only a world-view will enable the individual to make sense of scientific results that seem to run counter to deep-lying (religious, philosophical, scientific) beliefs.

Pauli's correspondence with the Swiss psychiatrist C. G. Jung – which began with his analysis (1931–4, by Erna Rosenbaum, one of Jung's disciples) and lasted until 1957 – reveals some stages of this fight. It is a rich and complex document, full of surprising and illuminating ideas. Long and detailed descriptions of dreams are used to explore areas incommensurable with natural science but eventually to be united with it. Pauli wrote in 1952:

> More and more it seems to me that the psychophysical problem is the key for the overall intellectual situation of the time; [we can advance by] finding a new ('neutral') unitary psychophysical language for

> describing an invisible potential reality that can only be guessed at by its effects, *in a symbolic way* [my emphasis].[25]

The key word is 'symbolic'. As used by Pauli it derives its meaning partly from quantum mechanics, partly from psychology. Quantum mechanics contains terms (the wave function) which seem to refer to things and processes but only serve to arrange phenomena in a systematic way.[26] In his Como Lecture which introduced the idea of complementarity, Bohr called such terms 'symbolic'. Physical objects are symbolic in an even stronger sense. They present themselves as ingredients of a coherent objective world. For classical physics and the parts of common sense dependent on it this was also their nature. Now, however, they only indicate what happens under particular and precisely restricted circumstances. Combining these two features Pauli envisaged a reality which cannot be directly described but can only be conveyed in an oblique and picturesque way. Making a similar point, Heisenberg wrote:

> Quantum theory is . . . a wonderful example for it shows that one can clearly understand a state of affairs and yet know that one can describe it only in images and similes.[27]

Psychology, long before, had led to analogous conclusions. There are events (apparently senseless actions, dreams etc.) which, taken by themselves, seem absurd but hint at causes different from their overt appearance. Might it not be possible, asks Pauli, to combine our new physics (matter) and psychology (mind, spirit) by means of characters, namely symbols which play a large role in myth, religion, poetry and thus to heal our fragmented culture?

It is clear that Pauli aimed at a new world-view. The fragmentation he wanted to overcome and the elements of a new unity even announced themselves in his dreams, in an emotionally charged manner. They were both represented by a magical figure, 'the stranger'. He was young, younger than Pauli. He was a wise man, a magician, conscious of his superiority, even over Pauli, contemptuous of his surroundings, especially of universities which for him were castles of oppression and which he tried to burn down. He spoke with force and conclusively. Women and children followed him and he began to teach them. Though opposed to science the stranger was tied to its terminology which means that he, too, was in need of salvation. Trying to get his message across he misused scientific terms in a systematic way. Such systematic 'misuse', Pauli found, is widespread; it occurs in the writings of engineers, laypeople and older thinkers. For Pauli this meant that there was a non-physical reality which, having been robbed of its language (rise of materialism) tried to make itself known in an indirect and 'symbolic' manner. Pauli thought that the exploration of this reality would be of importance for the 'Western Mind'.

## 6 What has been achieved?

Today many 'Western Minds' are enclosed in bodies that suffer from war, prejudice, illness, hunger and poverty. Everywhere in the world human beings face problems they cannot solve, not because they lack the right synthesis but because they don't have the power or, after years of suppression, even the will to act. Responding to so much suffering some intellectuals found new ways of making themselves useful. Quantum mechanics and the subtle problems it raises played no role whatsoever in their efforts. Medicine, a new and revolutionary form of Christianity, ecological concerns did; so did the belief that even the most downtrodden individuals know more about reality than was so far credited to them and the corresponding belief that Nature is open to many approaches. We have here a world-view that is not just an intellectual luxury, a kind of delicious dessert one consumes when the trivial matters of daily existence have been dealt with; it is a much needed response to some of the most pressing problems of our times. I mention this to impress upon you that the relation between quantum mechanics and reality is anything but a universal problem. It is not even a problem for all physicists; it is a problem for a limited group of rather nervous people who assume that their intellectual pains are felt all over the globe.

Wolfgang Pauli's efforts are not therefore irrelevant. True, he did not write for 'mankind' (though occasionally he spoke as if he did). Like the field workers in development, public health, primary environmental care, spiritual guidance, his efforts were for people he knew, his fellow intellectuals who, he thought, might experience problems similar to his own. In a way he even wrote for fighters like Frantz Fanon who was an intellectual and a psychiatrist and who objected to a purely mechanical revival of traditions.

> Such a revival can only give us mummified fragments which because they are static are in fact symbols of negation and outworn contrivances. Culture [a world-view] has never the translucidity of custom [established ideology]; it abhors all simplification. In its essence it is opposed to custom because custom is always the deterioration of culture. The desire to attach oneself to tradition or to bring abandoned traditions to life again does not only mean going against the current of history but also opposing one's own people.[28]

Fanon criticized African intellectuals who were fascinated by Western ways (forms of poetry, for example), who felt guilty, thought they had to do something for their own culture and started wearing traditional clothes and reviving old customs. But his criticism applies also to other areas. For example, it applies to more recent attempts, in the United States and elsewhere, to replace research by 'politically correct' cultural menus, it applies to the rhetoric of a New Age and to those realists who, being faced by the problems of the quantum theory eschew philosophy but continue to repeat realist slogans. Such actions, says Fanon, do not give us a culture or, as we might say, they do not give us a world-view, something we can live with. They 'not only go against the current of history, they also oppose the people one wants to inform'. Pauli criticized precisely such tendencies and he tried to overcome them in his letters and in his personal life. True, he looked at the past and he wanted to revive some of its aspects. But connecting the revival with new discoveries and keeping things in flux, he may have contributed to the emergence of a limited, but humane culture or world-view.

## Notes

1. E. O. Wilson, *On Human Nature* (Harvard University Press, 1978), pp. 192f.
2. Jacques Monod, *Chance and Necessity* (Vintage Books, New York, 1972), p. 170 (passage in brackets from p. 169).
3. Wolfgang Pauli, letter to Markus Fierz (8 August 1948) in C. A. Maier

(ed.), *Wolfgang Pauli und C. G. Jung, Ein Briefwechsel 1932–1958* (Springer, Berlin, 1992).

4. Leon Festinger, *When Prophecy Fails* gives examples of the power of world-views and tries to explain why obstacles increase it.

   This property is very useful. New ideas are crude, unfinished, unfamiliar and ill-adapted to their natural and social surroundings. It is easy for opponents to prove their (empirical, logical, social) imperfection. Ideas need time to develop their advantages and strength to survive initial attacks. They must therefore appear in the form of world-views, not of theories, and their defenders must disregard prima facie conflicts with logic, evidence and accepted principles. Many scientists proceeded in this way (examples in Paul Feyerabend, 'Was heisst das – wissenschaftlich sein?' in *Grenzprobleme der Wissenschaften*, P. Feyerabend and Chr. Thomas (eds.) (Verlag der Fachvereine, Zürich, 1985), pp. 385ff).

   But the property is also quite dangerous – political history, the history of medical and biological fashions provide numerous examples. It needs tact, wisdom, judgement to stop in time and thus to prevent disasters. Science would fail without a (non-formalizable) sense for the right balance between boldness and caution.
5. Galileo Galilei, letter to Benedetto Castelli (21 December 1613).
6. On early Chinese views of the Universe see Fang Li Zhi, 'Notes on the Interface between Science and Religion' in *John Paul II on Science and Religion*, ed. R. J. Russell, W. R. Stoeger SJ and G. V. Coyne (Vatican Observatory Publications, 1990). For Aristotle see Kurt Lewin, 'Der Übergang vom Aristotelischen zum Galileischen Denken in Biologie und Psychologie', *Erkenntnis* Vol. 2, 1931. For Kepler see Norbert Herz, *Kepler's Astrologie* (Carl Gerold's Sohn, Vienna, 1885), esp. pp. 24f.
7. Aristotle, *De Partib. Animal.*, 663 b 27ff.
8. Isaac Newton, *Opticks* (reprinted Dover, New York, 1979), query 31. Historians and scientists for a considerable time tried to keep Newton's theological and alchemical writings out of sight. How this was done is described in R. Popkin, *The Third Force in 17th Century Thought* (E. J. Brill, Leiden, 1992), pp. 189–94. Newton, these scholars thought, was a scientist, his theology was an aberration, almost an obscenity that had no place in a balanced account of his life. How could a scientist of Newton's stature waste his time on nonsense like that? Today the question is different. How did it happen, asks Popkin, that one of the foremost antitrinitarian theologians of his age could be sidetracked into astronomical and physical research? R. S. Westfall, *Never at Rest* (Cambridge University Press, 1980), p. 875 ff. gives the location of relevant manuscripts and literature, Frank E. Manuel, *The Religion of Isaak Newton* (Oxford University Press, 1974), background and quotations. B. J. T. Dobbs, *The Janus Face of Genius* (Cambridge University Press, 1991) contains a fascinating and amply documented synthesis of Newton's religion, science, his historical research and his alchemy and shows the impact his scientific discoveries had on his overall world-view.

Bacon had advised researchers to keep philosophy and revelation separated (*Advancement of Learning*, S. E. Creighton (ed.) (New York, 1944), p. 5). This assumed two sources of knowledge, or two books, God's Words (the Bible) and God's Works (the Universe). The idea that Nature should be approached without prejudice was therefore in part religiously founded: you accept what God tells you, you do not impose your 'interpretations' on it. Newton upheld the separation, most likely for political reasons. When he was President of the Royal Society 'he banned anything remotely touching religion, even apologetics' (Manuel, *The Religion*, p. 30). He did make theological remarks, but he put them into scholia, queries, letters – not into his main argument. Altogether he thought and argued in each area (theology, science, alchemy) according to the standards of this area.

Leibniz' assumption (first paper of the *Leibniz–Clarke Correspondence*, H. G. Alexander (ed.) (Manchester University Press, 1956), pp. 11f.) that exceptions to natural laws do not indicate an imperfection of these laws but God's intention to 'supply the . . . wants of grace' are characteristic for the (counter-empirical) world-view I have sketched in the text. When uttered it was as unrealistic, from a purely empirical point of view, as the assumption that this world is the best of all possible worlds.

9. Thomas Aquinas, *Summa Theologiae*, Ia,8,4.
10. Ockham's objection is found in his *Scriptum in Librum Primum Sententiarum (Ordinatio), (Prolog. Distinctio 1)*, G. Gál and S. F. Brown (eds.) (Franciscan Institute, St Bonaventura University, New York State, 1967), p. 241. Ockham's epistemology of individuals is directly related to this theological critique.
11. The situation in the nineteenth century is described in Johann Theodore Merz, *A History of European Thought in the 19th Century* (reprinted in four volumes, Dover, New York, 1965). Christa Jungnickel and Russell McCormmach, *Intellectual Mastery of Nature* (two volumes, University of Chicago Press, 1986) give a more detailed account of the situation in physics from 1800 to 1925 in German-speaking countries. Daniel Kevles, *The Physicists* (Alfred Knopf, New York, 1978) describes the history of American Physics from the second third of the nineteenth century up to the 1970s. See also H. von Helmholtz, 'On the Relation of Natural Science to General Science' in his *Popular Lectures on Scientific Subjects* (First Series, Langman Greens, New York, 1898) and du Bois-Reymond, 'Über die Grenzen des Naturerkennens', in *Zwei Vorträge* (Leipzig, 1882). Du Bois-Reymond explores the limits of the assumption, declared by Helmholtz to be fundamental for the physical sciences, that explaining the phenomena of Nature means 'reducing them to unchangeable attractive or repulsive forces whose intensity depends on the distance' (*Über die Erhaltung der Kraft* (Ostwald's Klassiker der Exakten Naturwissenschaften Nr 1, Leipzig, 1915), p. 6). Here the gulf between 'appearance' and 'reality' becomes very noticeable. On the low empirical content of classical hydrodynamics and aerodynamics see L. Prandtl and O. G. Tietjens,

*Fundamentals of Hydro- and Aerodynamics* (Dover, New York, 1957), p. 3. For an opposing point of view see C. Truesdell, *Six Lectures on Modern Natural Philosophy* (Springer, Berlin, 1966), Chapters 1 and 2 as well as Chapter 4, p. 85, on Prandtl.

12. For Gnosticism see Hans Jonas, *The Gnostic Religion* (Beacon Press, Boston, 1958) and R. M. Grant, *Gnosticism and Early Christianity* (Harper Torch, 1966). Early views about the nature of Christ are discussed in Jaroslav Pelikan, *The Christian Tradition*, Vol. 1, *The Emergence of the Catholic Tradition* (University of Chicago Press, 1971), Chapters 4 to 6. A more popular account discussing also changes of pictorial representation is the same author's *Jesus through the Centuries* (Yale University Press, 1985).
13. Xenophanes is discussed in W. K. C. Guthrie, *A History of Greek Philosophy*, Vol. 1 (Cambridge University Press, 1962), Parmenides in Vol. 2 (1965) of the same work. For Xenophanes see also Chapter 2 of P. Feyerabend, *Farewell to Reason* (Verso, London, 1987), for Aristotle Chapter 1, Section 6. In physics the transition to antiparmenidean views is connected with the second law of thermodynamics, the discovery that the most interesting problems in classical mechanics do not lead to integrable systems and chaos theory: Ilya Prigogine, *From Being to Becoming* (Freeman, San Francisco, 1980).
14. H. Weyl, *Philosophy of Mathematics and Natural Science* (Princeton University Press, 1949), p. 116.
15. Albert Einstein, *Correspondance avec Michele Besso*, P. Speziali (ed.) (Hermann, Paris, 1979), p. 312, see also p. 292.
16. *Albert Einstein, Philosopher–Scientist*, P. A. Schilpp (ed.) (Open Court, La Salle IL, 1949), p. 669.
17. See Arthur Fine, *The Shaky Game* (University of Chicago Press, 1986), p. 93.
18. J. Bernstein, *Quartum Profiles* (Princeton University Press, 1991), p. 84. The Bell-matter is by no means settled. New results may be obtained in moving reference frames (see O. E. Rössler, 'Einstein completion of quantum mechanics made falsifiable' in W. H. Zurek (ed.), *Complexity, Entropy and the Physics of Information*, SFI Studies in the Sciences of Complexity, Vol. 3 (Addison Wesley, 1990), pp. 367ff. See also the sixth answer given by Gerard 't Hooft in the discussion section. But even if it were settled there would still be dreams, feelings, religious experiences on the one side and the immovable realism of many scientists on the other.
19. Section 5 contains formulations first used in my column 'Atoms and Consciousness' in *Common Knowledge* Vol. 1, No. 1, Spring 1992.
20. Pauli's observations on the meaningful misuse of physical concepts are contained in 'Hintergrundsphysik' in C. A. Maier (ed.), *Briefwechsel*, pp. 176ff.
21. Wolfgang Pauli, letter to Max Born (21 January 1951). This letter and the two that follow are quoted from Ch. P. Enz and K. V. Meyenn (eds.), *Wolfgang Pauli, Das Gewissen der Physik* (Vieweg, Braunschweig, 1988), pp. 15ff.

22. Wolfgang Pauli, 'The influence of archetypal ideas on the scientific ideas of Kepler', in C. G. Jung and Wolfgang Pauli, *The Interpretation of Nature and the Psyche* (Bollingen Series, New York, 1955), p. 171.
23. Wolfgang Pauli, *Physik und Erkenntnistheorie* (Vieweg, Braunschweig, 1984), pp. 108ff.
24. Ibid, pp. 111f.
25. Wolfgang Pauli, letter to C. G. Jung (1952), in C. A. Maier (ed.), *Briefwechsel.*
26. Wolfgang Pauli, *Physik und Erkenntnistheorie*, p. 15 (footnote).
27. Werner Heisenberg, *Der Teil und das Ganze* (R. Piper, Munich, 1969), p. 285.
28. Frantz Fanon, *The Wretched of the Earth* (Grove Press, New York, 1963), p. 224.


## Comments

1. I was asked to stop telling stories and to give a systematic account of what I am up to. I am sorry I cannot do that because I don't want to be superficial. A systematic account is an account that rests on accepted premisses and accepted principles of argument. But accepted why? And by whom? You may say that the principles should be rational, or in agreement with experience. But science often advanced by arguing against experience, and principles that seemed rational to one generation often were rejected by another, and with excellent reason.

   To use an example: an enormous amount of experience conflicted with the idea that there are 'eternal laws of nature'. Yet combined with a form of realism and a judicious handling of the experience – reality distinction the idea led to a situation that seems to confirm it. (I say 'seems' for there is Bas van Fraassen and there is Nancy Cartwright.) Thus it is quite possible that an unreasonable belief ('unreasonable' according to some 'systematic' criterion) leads to a situation that is regarded as an 'improvement' by the 'reasonable' people though the 'reasonable' people do not agree with the interpretation some 'unreasonable' people put on it: van Fraassen would agree that modern science, including modern biology etc., is better than the science of Galileo and Descartes though he shares neither the realistic interpretation of the results of the development nor the belief that started it. ('But where are the patterns [of Nature]?' asks Nancy Cartwright of those who like Feynman define explanation as fitting phenomena into the 'patterns of Nature' (Nancy Cartwright, *How the Laws of Physics Lie* (Oxford University Press, 1983), p. 19)). Does this mean that reasonable people should encourage people with unreasonable beliefs as this would give them more reasonable results to interpret? How would they proceed? Would they start practising Plato's art of the Great Lie? And how is the issue going to be decided? I say by widening the domain of discussion, i.e. by renouncing 'systematic' accounts which are a fake and by bringing in the historical facts I have just mentioned.

2. The issue between van Fraassen and the realists more and more seems to become a verbal issue. Scientists say there are electrons and that they behave in certain ways. Van Fraassen agrees, the realists agree. Scientists explain why they think there are electrons. Van Fraassen accepts their explanations, the realists accept their explanations. Scientists also explain the role electrons play in certain effects, for example in the conductivity of metals. He accepts their explanations, the realists accept their explanations. There may be some disagreements here and there, they concern the scientific details, they concern assumptions such as the sea of occupied states in the Dirac equation and so on – but these apparently are not the differences we are talking about. As far as I can see the difference lies in this: the realists want to say that scientific assumptions, high theory included, in addition to having the properties just described *are about real things* while van Fraassen does not want to say this. Now what is it that is being asserted by the statement that a theory is about reality? That depends on the definition of reality accepted. Any such definition asserts that there are not only the theories, but that there are other things as well. Now I would assume that van Fraassen certainly agrees that there are not only theories but also other things such as scientists, birds, experiments and so on. He would also agree that scientific statements are connected, in a very complicated way, to all sorts of non-statement-like things and that it may be possible to unravel some of these connections by more theorizing or by better experiments. Moreover, he would say that theories were accepted and are being used by the majority of scientists because the connections are as described. But he refuses, and it seems to me rightly so, to replace this complex description by a simple relation, between a theory and a not-further-articulated monster, 'reality'. Saying that there is this relation does not add anything to what we know; rather, it takes away from our detailed knowledge by making things appear more simple than they actually are.

Regarding inferences to the best explanation: if I am right in saying that science is not one thing but many then, given a certain domain of facts, there will not be one inference to the best explanation, there will be many and they will be in conflict with each other. Recently Daniel Koshland, editor of *Science* was asked if it would not be better to give the money spent on the human genome project to the homeless. 'What these people [the people who asked the question] don't realize' he replied, 'is that the homeless are disadvantaged' – i.e. the homeless are genetic cases and the genome project will help them. Sociologists come to very different conclusions. Using inferences to the best explanation in both areas we get conflicting realities unless we engage in a little *ad hoc* reduction and re-interpretation.

# 8 Interpretation of science; science as interpretation[1]

as C. van Fraassen

In this chapter I shall address the realism–empiricism opposition only obliquely. My focus will be on the question: to what extent can we conceive of science as representation? Both the natural sciences and the fine arts, throughout their respective histories, have been characterized as activities whose primary goal is representation. This view is now widely rejected in the philosophy of art. Science too is now often characterized as interpretation rather than representation of nature.

While Paul Feyerabend's early writings were crucially involved in the (re-)emergence of scientific realism, they were always somewhat subversive of that view as well. After all, the summaries and slogans of realism have often depicted science as straightforward representation of nature, while Feyerabend's papers had as perhaps their most salient theme the role of interpretation at all levels of scientific observation and theorizing.[2] Our contributions to this book are therefore not likely to be confrontational – all the more so because one of my main inspirations in this topic area has been Feyerabend's *Wissenschaft als Kunst*.

I want to explore three levels of interpretation: the interpretative elements in a work, the interpretations which these works admit, and finally our interpretation of the very activity in which these works are produced. By 'works' I mean here both works of art (including literature) and works of science, that is, scientific theories. Since the representation/interpretation contrast is far more salient with respect to art, I shall begin there.

## 1 In what sense does a representation represent?

Today's prevalence of admittedly non-representational art, allows for the opinion that art may at least have been representational in the past, before these new developments established themselves as art. But in fact we do not need to explore the idea of representation very far to see the pervasive presence of interpretation everywhere.

In Vasari's classic survey of Italian art from the thirteenth to the sixteenth centuries we find the view:

> painting is nothing more than the simple portrayal of all things alive in nature by means of design and color as nature herself produces them.[3]

But to what extent does this describe even the simplest examples? Let us take Plato's example of a painting or drawing of a bed. Suppose for simplicity that it is merely a line drawing; then the drawing is the projection of a solid geometric figure, from a certain perspective, onto a plane figure on the canvas. Logically speaking, such a geometric relation could also hold accidentally between a solid and some lines 'drawn' in sand by rivulets of water. Such an accidental matching would not be a case of representation, so the representational relationship must be one that is *established intentionally*.[4]

Secondly, even in this simple case the viewer must learn how to 'read' the picture – the use of projection is a convention for *coding* the data selected for representation. Vasari's 'as nature herself produces them' ignores the intentional character of representation, the choice involved with respect to perspective, and the coding convention which neither nature nor painting wears on its sleeve, so to speak. Continuing this exploration of the concept, we'll quickly find ourselves coming to its limits, where it shades into its opposite or at least its contrary.

## 2 What exactly is faithful representation?

There are many ways to evaluate any particular representation (Is it beautiful? Does it have socially redeeming features?), but the criterion most closely related to the aim of representation is *accuracy*. But firstly, accuracy with respect to what is depicted is a matter of degree. Secondly, what is depicted is invariably selected from what could be depicted. These two considerations are relatively independent: *completeness* – how much is included, and how much is left out – must be distinguished from accuracy in a narrower sense, namely, as fit between what is shown and the part or aspect selected for depiction.

The bed drawing does not have the same dimensions as the bed; presumably the artist did not select size as a characteristic to be depicted. The criterion of accuracy is indeed in some sense the most basic criterion, but it presupposes a context, in which the question of selectivity is already regarded as settled.[5] This selection is not arbitrary, it too can be evaluated as apt, perspicacious, lacking, or even biased.

So even at the most basic level, the concept of representation has a curious complexity. For when something is offered as representation, it is thereby subject to two basic criteria – the first, though straightforward enough, presupposing as given the selection of features rendered (hence only contextually applicable), the other importing some value ('from outside', so to say) to determine the features to be selected. Although we have not yet strayed beyond the most basic sort of example of representation, we have now found four distinct elements that enter this concept. Representation is intentional, and the choices involved include a coding (reading) convention, a selection of aspects, and a 'fitting' (accuracy) relation which admits of degrees, and is relative to the context established by the preceding elements. Not only contextual adequacy but also the other elements admit of 'more and less'. When the coding or the selection becomes too egregious or simplistic, representation turns into mere symbolization or perhaps simile, allegory, or even metaphor. It may even be a metaphor to say in such a case that the work is a representation.

### 3 Representation *as* versus representation *of*

The concept of representation is not empty, even in current application.[6] Certainly we have many examples of works of art that represent, such as paintings of the adoration of the mystical lamb, of an angel's annunciation to Mary, of a woman with a swan who was a god, of the Emperor Napoleon, of several men and women having lunch on the grass. They depict something, they are depictions *of* something. But is that the main point, is that the crucial thing to say about them?

These paintings are not merely depictions *of* this, that, or the other; unlike, say, a census report, they do not only select some items and then simply encode them – they represent their subject *as* something. The adoring are represented as devout, the women as undaunted, demure, aghast, or resisting, the men as arrogant or vulnerable. But we do face a 'meta-question' here: can the idea of *representation as thus or so* be conceptually accommodated under that of *representation of this or that*? I will argue that the answer is *No*. The obstacles to a reduction of the interpretative element in representation are similar and indeed related to the obstacles that have withstood, in philosophy of mind, all attempts to reduce the psychological to the behavioural or physical level. The mere idea of representation is too poor to tell the story of representational art, because it is too poor to tell the story of perceptual experience itself.[7]

At first sight, it may seem that to represent, for example, certain men as arrogant is simply a matter of selecting certain characteristics for representation – to select besides their hair colour and posture also their arrogance. But when attention is directed to how art (and also science) represents *as*, interpretation takes on a crucial role. The pristine simplicity of the idea of mere representation, in the paradigmatic sense of a geometric projection, is altogether lost.

Consider for a moment that painting of a luncheon on the grass. Let us agree, if only for the sake of argument, that the men are represented as arrogant. How is this achieved? Perhaps we have in this painting an exact representation – a certain geometric solid has been precisely projected onto the two-dimensional canvas, and the colours-as-seen have been correctly filled into the corresponding areas, up to a certain varying level of detail – of just those aspects which the painter selected for depiction. But what of the arrogance? Unfortunately *being* arrogant is not equatable with having any one particular set of physical characteristics, and *looking* arrogant is not universally equatable with any set of visual characteristics describable in terms of shape and colour. So how exactly have the men been represented as arrogant?

Could it be that there is a particular representational code, such as a lifted eyebrow to indicate scepticism, which here conveys the arrogance? Does this painting perhaps belong to an artistic tradition which uses a highly elaborate set of apparently naturalistic details as symbols with recognizable iconographies, as in the religious art of the Renaissance? Are these men shown in a conventional posture of arrogance, in the way that in Mediaeval religious art men and women are depicted in conventional postures of sorrow, supplication, anger, and so forth? The answer would seem to be No to all these questions, in the case of this particular painting.

Of course, the artist could not depict the men as arrogant if they could not, in similar circumstances, convey their arrogance to us via some set of depictable characteristics. But 'arrogant' (like e.g. 'complacent', 'offended', 'friendly', 'hypocritical') is an adjective of interpretation. Whether or not a certain action, posture, or facial expression counts as arrogance depends on the whole social, cultural, and historical context in which it appears. How does all that get into this plane figure filled with colours?

It doesn't. The artist succeeds not by accuracy of represented details that univocally express arrogance, but by creating or provoking the relevant impression in the viewer/reader addressed. To represent the men as arrogant, the artist must enable us to see those depicted men in the depicted situation as arrogant – from within our culture, at a certain point in time, and with our specific history. Success in this respect, however, rests not only on what he does or shows. For if he did the same thing with another audience or public (who encounters the work in a different social, cultural, or historical context) the resulting interpretation could be markedly different.

To conclude then: the problem with the concept of art as representation is not that we now have examples of 'non-representational' art. The problem is rather that *representation as such is too poor, too meagre a concept*, to allow us to say much about any art at all.

Let me add here a reference to literary theory, where related distinctions have achieved a high level of sophistication. In genuine – as opposed to 'genre' – literature the author does not tell the reader what is happening or why, but *shows* it. Yet in 'only' showing the action, the author inevitably relies heavily on the interpretative activity of the reader who 'sees' what is 'shown'. The author's effort, therefore, is concentrated not on dictating the reader's understanding but on constraining and controlling the spread of interpretations available to the reader. Theoretical concepts successfully brought to bear on this process include Umberto Eco's notion of the creation of the model reader, and his opposition of *open* to *closed* text (see Eco, 1979). A work is open if (or to the extent that) it does not dictate its own interpretation. The closed text will tell us that the hero would cleave to his love, loved he not honour more, that his violence comes from unswerving loyalty, untainted with cruelty or lust. The reader is not guided so comfortably in *Crime and Punishment* or *Parade's End.*

## 4 The parallel case for science

Art, therefore, is not mere representation of something, but crucially involves interpretation. But what of science? Does science, in contrast, have as its aim exactly the faithful representation which art cannot or does not mean to achieve? Its medium is language – including of course especially mathematical language. The body of science is a body of information, a putative description of what there is in the world, and even of what the world as a whole is like.

The criterion of accuracy is applicable, but it divides again into two, as we saw for representation in general. It is easy enough to say something true, impossible to say all that is true about a given subject. Selectivity in science is deliberate and purposeful, effectively constituting its own subject matter.[8] This selection is not inscrutable; it is subject to evaluation as well. We ask not only if a given science provides accurate information about the aspects it has selected for attention, but whether it has selected well, whether it answers all or many of the *important* questions. Just as before, such evaluation draws on values current or imposed in its context, for what is important in the welter of data that assails us is not 'written on the face of' the data, nor is it yet another datum among them.

In the light of our discussion of representation, we can compare a 'pure representation' view of science with a contrary view. Recalling the amendments which were necessary for the view of art as representation, we arrive at the contrary view, that interpretation is also always involved, and indeed, enters at several different levels. It is easy to imagine how a relatively conservative philosopher of science might respond:

> Precision, accuracy, univocity, invulnerability to deconstruction or alternative interpretation are evidently the very hallmarks of rigor in the sciences – which is perhaps, if you are right, the very reason why scientific texts won't be literary texts as well, and why science is not art.

But let us scrutinize the sort of representation science provides.

Newton represented the solar system, accurately in many respects – the respects which he selected for thematic presentation – but he represented it *as* (what we now call) a Newtonian mechanical system. Obviously he abstracted from the facts, but does this consist – when perfectly successful – simply in *deletion* of the 'unselected' aspects? If so, abstraction can presumably introduce no inaccuracy or falsehood – what it produces is the truth remaining after we ignore some of the truth to be found.[9] But this irenic account of what Newton did – what he called his induction, his rigorous derivation from the phenomena – is too simple and too comforting, too good to be true.

How the solar system appears to Newton's God, how it appears in the view from nowhere, is not how it appears to us. Attending to what does appear, and has appeared to us, can we apply the interpretative adjective 'Newtonian'? Newton showed us that we could, by constructing a mathematical model and showing that it provided an adequate representation of the solar system. God created the world, Newton represented it as a Newtonian mechanical system, and we saw that it was good. Later Einstein represented it as a relativistic mechanical system, and again we saw that it was good – this time even better. The conclusion to draw is that the phenomena, to the extent Newton knew them, admitted his sort of representation – allowed being *represented as* a Newtonian system – but did not dictate that. They could equally be represented as an Einsteinian, relativistic system. We can draw a parallel here to the work of a portrait painter: he or she represents the subjects as arrogant, or as complacent, and the fact is that their comportment, as displayed to him, allowed both interpretations.

But someone might object: a serious disanalogy between science and art can be pressed here, after admitting to a minor analogy. By viewing the works of, for example, the Impressionists, or the Fauvists, we might become enabled to see nature, and humanity, in a new way. Analogously, Newton showed us how to see nature in a new way. Certainly, the new way of seeing involves the application of an interpretative attribute – the fact is only that the phenomena (how nature has appeared to us) *admit* of being classified as the appearances of Newtonian systems. Newton was wrong only in thinking that the interpretation, of the facts found so far, was unique.

This admitted analogy – so the objection would go – must be followed by a much more important disanalogy. The viewer may react to the painting, by seeing the men on the grass as arrogant, or as complacent – the painting represents them as admitting both interpretative responses. This does not deny that if there was or had been a real situation as original, that way of painting it was not compelled, but only allowed or *admitted* along with other alternative possible renderings.

Thus there are two levels at which non-univocal interpretation enters the scenario. With regard to science, the real situation corresponds to the solar system, the way the situation appeared to the painter corresponds to the recorded celestial phenomena (the data), and the painting corresponds to Newton's model. So science, like art, interprets the phenomena, and not in a uniquely compelled way. *But science itself does not in turn admit of alternative rival interpretations.* While there is ambiguity in the painting, and crucially so, there is no ambiguity in the scientific model. And so while the literary text is an open text, the scientific text is closed.

The history of science, however, puts the lie to this story, and in successively more radical ways. Gravitation, the only force treated successfully by Newton himself, is a *central force*, with the centre supplied by the gravitating mass. In the eighteenth century, it was taken as a principle of mechanical modelling that all forces in nature are central forces. Was this an addition to Newton's science? We must first reflect on this question itself, and ask what kind of answer it requires. Does it ask whether Newton deliberately omitted the principle from the principles of mechanics, or whether he indicated it tacitly, so that it was there for him but as a principle which had not risen to the level of explicit formulation? All Newton's models are of the type admitted in the eighteenth century; it is as if he already had that principle as well. But 'as if' is all we can say. Should we instead take Newton's science to be defined solely by what was explicitly stated?

In that case all the Newtonians would appear to have misunderstood Newton's mechanics. For, to take one example, Laplace only formulated the common understanding when he used the dramatic device of an omniscient genie to convey that the Newtonian world picture is entirely *deterministic*. But if we look only to explicitly formulated principles, we must say that this science was not deterministic. The law of conservation of energy was not recognized as an independent and needed addition until the nineteenth century – perhaps partly because non-conservative systems had not been sufficiently well conceptualized – and the science allows for indeterminism before that is done.[10]

What retrenchment could come next? Newton managed to create in this audience (the physicists and educated lay public of the modern era) the impression of total determinism with such force, that their own view of science began to include it as a criterion – the *telos* of science is representation *as* deterministic. The task of science is not finished till that is done – and of course, except for details, it has been done – that is the implicit conviction of the nineteenth century, at least in the most visible quarters. It was not shared by all, C. S. Peirce being an honourable exception, for example, and indeed it was compelled neither by the phenomena, nor by the science, nor by its success. Science itself admitted different interpretations, at each stage, even if at each stage, one interpretation seemed to be dominant.

The admission of alternative interpretations is spectacularly visible in philosophy of physics today, with respect to quantum mechanics. The implications of interpretative questions there I shall take up below at some length. At this point we already have at least this conclusion: in science too, we find interpretation at two different levels: the theory represents the phenomena *as* thus or so, *and* that representation itself is subject to more than one tenable but significantly different interpretation. The texts of science too are open texts.

## 5 Spectres of vagueness and ambiguity

The representations that science and art present to us interpret as well as represent their subject matter, and succeed in part by evoking interpretations of themselves by us. Neither on canvas nor on the page does representation uniquely dictate how we are to understand it. We should now inquire into the role and importance of the multiplicity of interpretations which the work admits. Is this to be taken as a defect, an obstacle which the artist and/or scientist strives mightily to negate? Does the artist's or scientist's success consist in blinding us to all but a very narrow range of interpretations, and thus determining our interpretative reactions? Or is there rather a special value or virtue to be found through this interpretational multiplicity?

I will leave art aside from here on, and simply direct these questions to science. When more than one reading is left open, a representation is first of all *vague*: the information conveyed is at best a disjunction of what it says on the various, diverse possible readings. In addition, it may be *ambiguous*: there may be a certain tension between its various salient meanings, they may even be at war with each other, so to speak.

That the scientific representation of nature may harbour ambiguity as well as vagueness came sharply to the world's attention in the early history of the quantum theory. It seemed at first that this theory provided us with two rival, incompatible, 'complementary' pictures of the micro-world, as consisting of particles or of waves. Each picture was to be mobilized in reading the same formulas, when applied to different experimental contexts. This complementarity was seen by Bohr and a number of subsequent writers as the transcendental clue to the interpretation of quantum theory, or even of the world itself. Others saw here the demise of the enterprise of representation altogether. Perhaps representation, or even interpretation, was after all a merely anthropocentric aim, restricted to heuristic associations of pictures with strictly speaking uninterpretable theoretical and symbolic devices.

Radical conclusions from the idea of complementarity do not any longer enjoy the high regard they once had. Scientists educated in classical physics had two sorts of pictures, two sorts of models, which were mutually exclusive: the wave picture and the particle picture. Different processes were modelled in these two different ways, and no process could be of both sorts. For a while in the twentieth century, scientists were using both sorts of pictures for the same processes, though in connection with different experimental set-ups. Sometimes the behaviour of light, for example, admitted representation as a wave in a medium and sometimes it admitted representation as a stream of particles. Bohr's quite revolutionary idea was that this could be accepted as a normal and satisfactory state of affairs, that a theory could simply offer two families of models, with some prescription about when to switch from the use of one to the use of another. The idea was workable only, however, if that prescription itself was not equivocal, and hence only if the scientist's apparatus could be exempted from this, and could be said to have a univocal description. But the only univocal description available was that of classical physics, which unfortunately predicted wrongly even at the macroscopic level of the apparatus. Today there seems little hope of re-instituting complementarity as the key to interpretation of physics.

While this worry seems today rather out of date with respect to our understanding of quantum theory, the crises did not cease when the idea of complementarity was relegated to its proper, rather less fundamental, place. Another very basic tension which brought along vacillation, and hence suspicion that the theory cannot have a satisfactory interpretation, was perhaps first made fully explicit in Wheeler's commentary on Everett.[11] On the one hand quantum mechanics is putatively the fundamental science, in principle encompassing all sciences as parts, and in principle affording a complete description of the world. On the other hand, much of it developed in the form of a theory of partial systems – systems studied in relation to an environment, in terms of input from and output to that environment. The question is then whether the title of E. B. Davies' book *The Quantum Mechanics of Open Systems* really describes the entire theory (with every aspect of an environment being potentially part of a described system, which will itself however always be described as open to an environment), merely a sub-theory, or a proper extension of the theory. Interpretations of quantum mechanics presently available differ on this question; and even when they agree, they differ in other significant ways.

However, the question about value of ambiguity should be subdivided into two. First we ask whether ambiguity and openness have been of value to science in practice, or instead have hampered its progress. Then we must ask how different philosophical views rule on whether ambiguity and openness are defects, or alternatively, can be valuable to science.

As to the first question, no philosopher should prejudge the history, sociology, or psychology of science. At every point in the history we see both blindness and insight, and the two are inseparable. The insight that Newtonian mechanics lent itself to being the mainstay of a deterministic world-view, blinded the Enlightenment to the possibilities of indeterminism. Prima facie, at least both that insight and its correlate blindness are to be credited with inspiring the spirit of research which led to such triumphs, and *also* to the phenomenal limits, of classical physics. But then the previously unseen alternatives – the previously undetected gaps, vaguenesses, and ambiguities – became visible as it was realized that science did already have resources to begin the study of discontinuity and chance in nature.

But the second question, how the presence of vagueness or ambiguity in science must look *sub specie* different philosophical views of science, is more probing. Philosophical views of science are interpretations of this, to us very salient and indeed more and more pervasive, aspect of our civilization. Such interpretation of science takes its place at one further remove from nature, after scientific theories interpret as well as represent the phenomena, and those theories themselves are subjected to interpretation in foundational enquiry. Prominent in our discussions at the symposium were the two opposed views of the aim of science which I call empiricism and scientific realism. The empiricist view is that scientific theories need not be true (overall) to be good, they need only be accurate in their representation of the phenomena, and how they interpret those phenomena is of value only in the pursuit of this primary good, i.e. empirical adequacy. The scientific realist view is instead that the defining aim of science lies with truth: the aim is to provide us with a true story of what is really going on, behind the scenes as well as on the stage, so to speak.

There are nuances, of course. A more or less naïve scientific realist (not likely to exist in reality, of course) would have to say that ambiguity, vagueness, and gaps are all defects. The last two spell incompleteness of achievement with respect to the literally true story of the world. The first also obstructs, sabotages, such achievement as it drives thought into several different directions at once – reaching for the aimed-for achievement continues only with the elimination of conflicting interpretations.

But what if the empirical predictions remain invariant under all ways of resolving the ambiguity (all ways of opting for one interpretative completion over its rivals)? Then empiricism at least sees no defect. Indeed, the only true way to enhance our understanding of science, for the empiricist, is not to resolve such ambiguity, but to find out in how many different ways it could be resolved. Each tenable interpretation will throw new light on the theory, by showing that this is how the world could be as the theory describes it. Any such new light is valuable. Since each of those ways of seeing the world is potentially a good way to respond to new phenomena as yet unexpected or even unimagined, they are not to be chosen among, but valued and appreciated. Indeed, the tensions created by ambiguity, like the paradoxes about infinity and infinitesimals that plagued modern mathematics, may well be the crucial clues to creative development.

## 6 Interpretation of quantum mechanics

As a concrete case, I will now discuss some recent work in the interpretation of quantum mechanics, subsequent to the episodes related in the previous section.

An interpretation of a theory is an attempt to answer questions which that theory left open.[12] This presupposes that some are left open. To check whether that is so, presupposes in turn that we can identify *the theory* (the 'official' theory).[13] We recognize that the division is to some extent arbitrary. The models provided for us by the theory are the Hilbert space models equipped with Hermitean operators to represent observables and statistical operators ('density matrices') to represent states. This leaves some indefiniteness, for which of the Hermitean operators do represent observables? All, or only some? What superselection rules may be introduced? (When there are non-trivial superselection rules, then not all Hermitean operators represent observables, so these questions are connected.) Even in the case of the Schrödinger equation there is some indefiniteness: must $H$ be an observable? This is connected with the preceding two questions: it amounts to the question whether $H$ commutes with all superselection operators (in which case dynamic evolution cannot move a system out of one coherent subspace into another).

These questions may be among the more abstruse, but everyone agrees, I think, that some topics were rightly controversial and subject to dispute from the beginning: what exactly happens in a measurement, the physical significance of distant correlations, and the extent to which the 'correspondence principle' bestows a classical character on nature at the macroscopic level, to mention but three of the most obvious.

But granted that there are open questions, and hence interpretation is needed, what *attitude are we to take to this*? One possible attitude is to maintain the conviction: there must be a unique right interpretation. It is in general not easy to give a verdict on the consequences of such an attitude or its rivals. In any such inquiry as this one, the criteria of adequacy are not likely to be entirely clear and explicit at the beginning; they may well emerge only along the way. But in this case, we do not stand at the beginning. One great difference between now and fifty years ago is that *now*, if someone proposes an interpretation of quantum theory, they will face a large set of stony-faced hard-headed critics, with a real arsenal of possible objections and difficulties. Today, specifically, the quantum theory of measurement is so well-developed that no interpretation can survive unless it obeys very high standards of rigor.

Yet there are a number of rival interpretations under study today which appear to have a good chance of surviving. We can divide them into two groups. The first group is in agreement with von Neumann's interpretation concerning the 'Eigenstate–Eigenvalue Link'. That means: an observable $A$ pertaining to a given system $X$ has a specific value (one of its eigenvalues) if and only if $X$ is in an eigenstate of $A$. The second group consists of interpretations in which this link is broken in at least some cases.

In the first group we find of course von Neumann's own interpretation, with its notorious 'acausal' transitions (collapse of the wave packet, Projection Postulate). There are also two other types in this group, which explain those transitions in some way (or explain them away, if you like). One of them sees the clue to the 'collapse' in thermodynamics (Danieri, Loinger, and Prosperi; Hepp). The other finds the clue in superselection rules (Kai-Kong Wan; Beltrametti and Cassinelli). I do not include here the GRW model (Ghirardi, Rimini, Weber) because, although they accept collapses, they also accept an indeterministic 'swerve' in Schrödinger's equation with calculable (though extremely small) observable effects which diverge from those predicted by the quantum theory.

In the second group, the interpretations deny the Eigenstate–Eigenvalue Link. In other words, they allow that an observable may have a value when the system's state is not an eigenstate of that observable. Since an extra parameter is needed to specify that value (extra in addition to the state, that is), these are all among the 'hidden variable' interpretations. Examples are:

(1) The Ensemble interpretation (Rosen, Popper, Ballantine)
(2) Pilot Wave (de Broglie–Bohm)
(3) 'De-occamizing' (this is Redhead's name for the sort of interpretation in which one operator can stand for more than one observable; Gudder, the 'anti-Copenhagen variant' of the Modal interpretation)
(4) Many Worlds (Everett, De Witt)
(5) Interaction algebra (Kochen, Bub)

Some of these we can already be quite sure do not work; for example the versions of the Ensemble interpretation apparently favoured by Rosen and by Popper. All of them have counter-intuitive consequences, which may or may not threaten to turn into defects that will eventually show them to be untenable after all. The Pilot Wave interpretations, for example, are surprisingly resilient, but they may be incapable of yielding relativistically invariant pictures of what is going on in the micro-world.

In addition to the above there are the Modal interpretations, also in this group; for example those devised by Healey and by Dieks, and also my favorite, the 'Copenhagen variant' of the Modal interpretation. Needless to say, these too have their counter-intuitive (and possibly problematic) aspects.

In general, we cannot expect interpretation to be unique. If more than one genuinely mutually independent question is left open, and the theory is not inconsistent, then necessarily, there will be more than one interpretation.[14]

Among philosophers especially I have encountered the attitude that we don't really understand a theory if there is any point when we don't know what to say. And if any question can be answered in more than one way, i.e. if there is more than one *tenable* interpretation, then indeed, objectively, we don't know what to say.

But I think this is a wrong view of understanding. It misunderstands what it is to understand. Every time we discover a new tenable interpretation, we understand more, we understand the theory better. Even if we find it is tenable only by some criteria, or if eventually it fails completely, we have gained a good deal more understanding from that. What we have seen then are possible ways the world could or could not be if the world is as the theory says it is.

I submit respectfully that the quantum theory admits of a whole spectrum of diverse, tenable interpretations, and that this is a healthy (and in principle, characteristic) state for science to be in. Every discovery of a tenable view increases our understanding. Each time we find that an interpretation is not tenable, that reveals constraints which limit the range of tenable interpretation, and *thereby* also increases our understanding. Tolerance of ambiguity is a great virtue for philosophers as well as for scientists, and helps rather than hinders understanding.

Both when a new tenable interpretation is found, and when a putatively tenable interpretation is shown wanting, the most obvious benefit is the refutation of common claims and received wisdom. Consider for example the claim that quantum theory implies that particles cannot have a definite spatial trajectory. Bohm's interpretation, despite some objections we may bring against it elsewhere, clearly shows at least that this is not logically implied by the theory. To discover that, after hearing the contrary claim so often, certainly makes for an increase in understanding.

Arguments to the effect that one or other *tenable* interpretation must be or cannot be **right** or **true**, can only be based on considerations from outside of physics. Such arguments therefore cannot increase our understanding of quantum theory, though they might reveal something about relationships between physics and something other than physics (perhaps somebody's preferred metaphysics).

## Notes

1. This paper overlaps parts of van Fraassen and Sigman (1992) and van Fraassen (1991a).
2. Consider especially the famous Thesis 1 of the 'pragmatic theory of observation' in Feyerabend (1958).
3. Cited in Gombrich (1960), p. 12.
4. Compare the beginning of Putnam's 'Brains in a vat' in Putnam (1981).
5. For both accuracy and selectivity, one could ask whether the limiting case is in fact possible, that is, perfect accuracy of depiction with respect to some selected feature, or total completeness of selection of features for depiction. But the existence or possibility of the limiting case is not required for evaluation of more and less to be possible. For example, there is no longest distance; however, the possibility of comparing distances gives the concept of distance its legitimacy.
6. The vagueness in the concept implied by our observations so far does not suffice in any way to deconstruct the concept. Slippery slopes and continua are characteristic of almost all our concepts in use; that does not remove the reality or non-triviality of the distinctions they embody.
7. This, of course, is an area of philosophy of mind in which issues have been hotly contested of late. See Lynne Rudder Baker (1987), and Hilary Putnam (1988), especially his chapter 'Why functionalism didn't work'.
8. This has especially been emphasized in writings on science in the phenomenological tradition; see e.g. Kockelmans (1962, 1987).
9. This view of abstraction is quite prevalent, even if sometimes disputed. It appears to be strikingly refuted in statistical representation by Simpson's paradox (see van Fraassen (1988) and (1991), chapter 8, section 5).
10. See Earman (1986), and further van Fraassen (1991), chapter 10, section 4.
11. Contrasting one specific formulation of these two ways of viewing quantum mechanics, John Wheeler wrote, 'Our conclusions can be stated very briefly: (1) The conceptual scheme of "relative state" quantum mechanics is completely different from the conceptual scheme of the conventional "external observation" form of quantum mechanics and (2) The conclusions from the new treatment correspond completely in familiar cases to the conclusions from the usual analysis.' (Wheeler (1957), p. 463.) This contrast is not essentially bound up with the specific account (Everett (1957)) on which Wheeler was commenting.
12. Recall the levels of interpretation: (1) setting the various parameters of representation, such as selection of the aspects of the phenomena that are to be represented, is an element in the interpretation of nature; (2) adopting one of the possible 'readings' of the representation when we

reflect on the theory/work is a second element of interpretation; (3) adopting a view of the activity – e.g. to what extent its aim is representation, criteria of adequacy – adds a third level of interpretation. In this section, I am discussing moves made on the second level, and my discussion is likely to be biased by my empiricist (third level, in this hierarchy) view of what science is all about.

13. That is not always easy, though in the case of quantum theory it is facilitated by the widespread acceptance of John von Neumann's book as an authoritative exposition of the mathematical foundations of the theory. Even here the assumption that we know how to separate out von Neumann's own interpretation is a bit fragile, despite his conscious effort to demarcate it in his text.
14. A number of the interpretations listed here are discussed in van Fraassen (1991); see also van Fraassen (forthcoming) and van Fraassen and Sigman (1992).

## Bibliography

Baker, L. R. 1987. *Saving Belief.* Princeton University Press.

Bohm, D. 1952. 'A suggested interpretation of the quantum theory in terms of "hidden variables"'. *Physical Review* **85**, 166–79 and 180–93.

Dieks, D. 1989. 'Quantum mechanics without the projection postulate and its realistic interpretation'. *Foundations of Physics* **19**, 1397–1423.

Earman, J. 1986. *A Primer on Determinism.* Dordrecht: Reidel.

Eco, U. 1979. *The Role of the Reader: explorations in the semiotics of texts.* Bloomington, IN: Indiana University Press.

Everett, H. 1957. '"Relative state" formulation of quantum mechanics'. *Reviews of Modern Physics* **29**, 454–62.

Feyerabend, P. 1958. 'An attempt at a realistic interpretation of experience'. *Proceedings of the Aristotelian Society.* Reprinted in Feyerabend (1981).

—1981. *Realism, Rationalism and Scientific Method. Philosophical Papers,* vol. 1, Cambridge University Press.

Gombrich, E. H. 1960. *Art and Illusion.* Princeton University Press.

Kockelmans, J. 1962. *Phaenomenologie en Natuurwetenschap.* Haarlem: Bohn.

—1987. 'On the problem of truth in the sciences', *Proceedings and Addresses of The American Philosophical Association* **61**, #1 (Supplement), 5–26.

Putnam, H. 1981. *Reason, Truth and History.* Cambridge University Press.

—1988. *Representation and Reality.* Cambridge, MA: MIT Press.

van Fraassen, B. 1988. 'La città invisibile/The invisible city'. L. Mazza (ed.), *Catalogue of the Milan Triennale Exhibition.* Milan: Electa.

—1991. *Quantum Mechanics: An Empiricist View.* Oxford University Press.

—1991a. 'La meccanica quantistica: uno spettro di interpretazioni' *Iride* **7**, 28–50.

—'Interpretation of quantum mechanics: parallels and choices' forthcoming in a volume edited by L. Accardi.

van Fraassen, B. and Sigman, J. 1992. 'Interpretation in science and in the arts'. G. Levine (ed.), *Realism and Representation*. University of Wisconsin Press.

J. Wheeler 1957. 'Assessment of Everett's "relative state" formulation of quantum theory'. *Review of Modern Physics* **29**, 463–5.

# 9 Problems in debates about physics and religion

Willem B. Drees

## 1 Introduction

It has been said that 'scientific realism is a majority opinion whose advocates are so divided as to appear a minority' (Leplin 1984, p. 1). Something similar happens in the reflections on physics and theology in this volume: even though most contributors share a positive attitude towards religion, they are strongly divided. The discussion is not merely one about answers, for instance whether or not the Universe had a beginning, and whether that supports belief in a divine 'first cause'. The real disagreements concern the questions, the way the problem is posed. Failures to communicate may have their roots in that which is considered obvious, and thus often left implicit, rather than in that which is the explicit topic of the debate. Thus, the aim of this chapter is to achieve a clearer view of the variety of approaches in relating physics to religion.

Before embarking on the main topic, I will briefly argue for the relevance of such reflections.

There is a public relations problem for theology in an age of science. Affiliation with churches is declining. 'Belief in God' is considered by many to be an option which is superfluous, outdated, falsified, meaningless, or a matter of private taste. The role of science in the rise of such attitudes should not be overestimated; many other social factors have also contributed. However, clarifying the relation of religion and science seems important to the future of religion.

There is also a public relations problem for science. Among those for whom religion is important, some feel threatened by science. A

Gallup Poll in November 1991 found that 47 per cent of all Americans supported a strict creationist view ('God created man pretty much in his present form at one time within the last 10 000 years'), whereas 40 per cent endorsed a religious view which had adapted to evolution ('Man has developed over millions of years from less advanced forms of life, but God guided this process, including man's creation'), and only 9 per cent endorsed a naturalist view ('Man has developed over millions of years from less advanced forms of life. God had no part in this process').[1] Defusing feelings of threat, and hence an anti-science attitude, may well be a relevant PR motive for scientists to participate in dialogues on science and religion.

Antagonism is one attitude. Unrealistic expectations with respect to science are another social phenomenon which might be a reason for concern. For instance, a Natural Law Party participated in British elections in April 1992. They claim that superstrings and supergravity confirm old Vedic insights, as presented in the West by the Maharishi Mahesh Yogi. Many more examples of misguided spiritual or material expectations could be listed. Though such expectations may contribute to a friendly environment for the funding of science, they may do harm as well, as they may give rise to counter-reactions when expectations are unfulfilled. Besides, they inhibit the critical role of science through their quest for an encompassing view.

Even though public relations are important, utilitarian grounds should not carry the whole weight of the science-and-religion interaction. In my view, three less pragmatic motives are essential as well. Firstly, there is the need for honesty which presses one to search for consistency, also in matters of religion. Secondly, curiosity may well drive one to relate various fairly distinct human enterprises, since new ideas and insights might come up in an encounter. And, thirdly, concern about the well-being of humans and other living things may be a reason to reflect upon the impact of science and of religion on our self-image and on our culture.

The next section will discuss various views of the problem by making a brief historical tour (2.1) and by arguing that our situation differs from the mediaeval synthesis of religion and science not merely with respect to the content of science, but also in our view of knowledge and in our appreciation of the world (2.2). These three types of differences may be related to three views of the 'object' of religion: God, mystery or meaning. This leads to a variety of interactions with science (3). Starting from the scientific side, naturalistic and monistic tendencies related to science provide an important context for reflection on religion (4). In the concluding section (5), various niches for religion will be discussed.

## 2 Various views of the problem

### 2.1 A brief history of relevance and separation

In the Middle Ages religion and pre-scientific knowledge were integrated. The rise of modern science led to conflicts. This resulted in modern atheism as well as in religion withdrawing from the cognitive domain to apparently inaccessible realms such as ethics or feelings. This story will be sketched and nuanced below. The issue then becomes in what way religion may regain a place in the cognitive realm, or whether it should develop more in line with the insights which resulted in separation.

#### *The medieval synthesis*

A synthesis of religious convictions and (pre-) scientific insights seems to characterize the late Middle Ages. Prominent among the scientific insights were those of Aristotle, as mediated through Arabic culture. A major example of a theological system in coherence with the available knowledge is the work of Thomas Aquinas (thirteenth century). However, the synthetic approach has not been restricted to systematic theology. Fiction, like Dante's trilogy about heaven, hell, and purgatory, illustrates the medieval quest for a world-view as well. Ideas taken from the Greek philosophers (Plato, Aristotle), from Holy Scripture, and from the writings of theologians of the first millennium were integrated.

Among the characteristics of the medieval synthesis were its static character, its hierarchical structure and its geocentrism. The order, built upon Aristotle's doctrine of 'natural place' was not merely understood as something factual; the order was also prescriptive (as some uses of the words 'natural' and 'counter-natural' in our time still reflect). The order of the world was not something accidental, but reflected God's order and providence.

Though some elements in the old synthesis were lost with the rise of modern science, especially its geocentrism (and thus its Aristotelian doctrine of 'natural place'), at first most scientists argued more or less in the same way with respect to religion. The order of the world as science discovered it was the order God had put into it. One could learn about God from the book of Scripture and from the book of Nature. It is a misunderstanding that the beginning of modern science was marked by conflicts between science and religion, or between scientists and the Church. To make clear that the situation has been more complex, let me make a brief digression on the Galileo affair.

### *The Galileo affair: relevance or neutrality?*[2]

The conflict that arose in the seventeenth century around Galileo has often been seen as a conflict between science and the Catholic Church, or even theology in general. However, the conflict also reflected a clash between two views of, and social contexts for, science. There was the scholastic tradition, appealing to previous authorities (Aristotle as 'The Philosopher') and well established in the medieval universities. The new sciences arose in another setting, in combination with trade and crafts in the cities which were gaining in importance. Whereas the then traditional approach equated knowledge and certainty, the new approach led to modern empirical science, ascribing a more provisional and probabilistic status to knowledge. The Galileo conflict may be seen as marked by a specific alliance between the medieval, scholastic tradition in knowledge and certain powerful elements in the Catholic Church rather than as a straightforward conflict between conservative religion and progressive science.

The Galileo affair may also be described as a conflict between two views of religion, especially two views of the relevance of Scripture for science. In Galileo's writings, for instance his Letter to Grand Duchess Christina (1615),[3] two types of argument about the proper way of dealing with the relation between Scripture and natural science can be found.

(1) *Relevance.* If science has proven certain facts, one has to adapt one's interpretation of Scripture. However, if scientific knowledge is merely 'plausible opinion and probable conjecture' in place of sure and demonstrated knowledge, one would have to give priority to Scripture. 'Where human reasoning cannot reach – and where consequently we can have no science but only opinion and faith – it is necessary in piety to comply absolutely with the strict sense of Scripture.' Thus, the information of Scripture is relevant to our view of the world and vice versa, but Scripture or science needs re-interpretation if a conflict arises. Which one needs re-interpretation depends on the certainty of the claims made by both sides. By the way, science may be open to re-interpretation in two ways. Either the substance itself may be interpreted differently, or the status of scientific statements may be assessed differently. For example, Copernicus' book *De Revolutionibus Orbium Caelestium* (1543) had an anonymous preface (ascribed to Copernicus but now known to have been written by the Lutheran theologian Osiander) which emphasized that these ideas were no more than hypotheses, developed for simplifying calculations, and did not aspire to be true.

(2) *Neutrality.* Galileo also argues for the neutrality of Scripture in matters cosmological, and vice versa. He quotes an ecclesiastical authority who states 'that the intention of the Holy Ghost is to teach us how one goes to heaven, not how heaven goes'. The Bible is only relevant in matters 'which concern salvation and the establishment of our faith'.

### *Relevance and neutrality*

After the rise of modern science various positions co-existed, of which some assumed 'relevance', whereas others took off from 'neutrality'. Some argued for a restricted scope of science and of religion, which might allow co-existence through mutual neutrality. Others argued that science and religion cover more or less the same issues (everything?), which implies relevance of the one for the other, either negatively (conflict) or positively (harmony).

An antagonistic attitude, maintaining an older view of the world despite more recent scientific discoveries, is an example of relevance. Such conflicts continue until our time, 'creationism' being its most well-known contemporary manifestation. Similar to this rejection of science for religious reasons is the rejection of religion as a consequence of scientific discoveries. Characteristic titles in this context have been A. D. White, *A History of the Warfare of Science with Theology in Christendom* (1896) and J. W. Draper, *History of the Conflict between Religion and Science* (1875). However, as a balanced view of history, this 'warfare' model is far from being the whole story. It too is to be seen in its social context (e.g. Brooke 1991, pp. 33–42).

Others adapted to new discoveries, arguing for the meaningfulness of religion in terms offered by the sciences. This may be exemplified by the English 'arguments from design' tradition. If one were to find a complex item on the shore, and it turned out that that item could be interpreted as indicating the time (a watch), one could argue that the item was designed for that purpose, rather than that the correlation between the position of the Sun and of the hands on the clock was a mere coincidence. Similarly, so the argument goes, one should opt for intentional design of organisms, as their intricacy surpasses the intricacy of watches by far.

*Neutrality* has expressed itself in emphasizing the differences between religion and science. The idea that religious convictions are neutral with respect to scientific ideas has continued to attract major thinkers over the centuries. One well known example is the philosopher Immanuel Kant, who discussed in his *Kritik der reinen Vernunft* (1781) the status and limitations of theoretical, scientific knowledge, whereas he introduced religion (God, soul, immortality) in the context of his *Kritik der praktischen Vernunft* (1788), his major work on ethics. The theologian Friedrich Schleiermacher (*c.* 1800) placed the feeling of absolute dependence at the core of religious life, thus incorporating the neutrality principle. Feelings or attitudes may also be a meeting ground for science and faith. For instance, the historian of science Olaf Pedersen has argued that, 'when a scientist realises the implications for one's personal existence of the fundamental scientific experience, he has adopted a relationship towards the world which is essentially the same as that which the believer adopts when expressing belief in creation' (Pedersen 1988, p. 138). Science does not disclose God's attributes. Rather, science and faith have to do with similar attitudes towards the world.

A more recent defence of neutrality is provided by the philosopher Ludwig Wittgenstein, who in his later writings emphasized that the meaning of words is to be understood through their use. Thus, the meaning of the term 'goal' may be clear in the context of football, but it is unintelligible in the context of chess. Various practices constitute various language games. Transferring concepts from one language game to another language game, say from religion to theoretical physics or vice versa, is considered a misguided enterprise.

The history of interactions of science and religion does not provide us with a single coherent view of the nature of the problem. Some suggest a lack of adaptation of religion to scientific insights (e.g. Wildiers), whereas others suggest that Christian apologetics went wrong by adapting too much to the terms of the discussion as set by the natural sciences (e.g. Buckley, Dillenberger). Historians, and others, describe the interaction of, or confrontation between, science and religion in terms which fit their own agenda. That is not merely an agenda with respect to the relation between science and religion; their own view of science, religion and human culture is involved. Science-and-religion is not comparable to building a bridge between two territories which are sufficiently well-known. Rather, in reflecting on the connection, one also develops and adapts one's view of the nature of religion and the scope of science.

Besides, in relation to realism in science as discussed in other essays in this volume, it may be noted that those who defend religion are not all on the same side in the realism–empiricism debate. Playing down the claims and scope of science may be one way to create room for religion (Buckley, 1987, 1988; see also Feyerabend, Hesse and van Fraassen in this volume), but scientific and theological realism may also be aligned (Wildiers, 1982, see also Davies in this volume).

### 2.2 Three 'variables'

*New knowledge* separates us from the medieval synthesis. Among the changes have been the loss of geocentrism, the longer time scales, and the evolutionary understanding of humanity in relation to the rest of the living world. However, there are at least two other types of change that should be considered as well.

It was not only the knowledge of Nature that changed, but also our ideas about *the nature of knowledge*. The *subject* acquired a more prominent place. For instance, the philosopher Kant understood as inaccessible the world as it is in itself; the accessible world is the world as *we* describe it in terms of our categories. Although specifying these categories has turned out not to be a straightforward matter, the insight still stands that knowledge is shaped by our categories and not just by the reality it intends to be about. The emphasis on the subject, with his or her categories of thought, is continued in our century with the emphasis on the role of language and context and with the decline of belief in foundations which would provide for certainty. Such issues are central to other essays in this volume: in what sense is our knowledge knowledge *about the world* (realism) and to what extent is it *our construction*, relevant in a specific practical context but not to be granted a more universal meaning independent of that context?

A third change regards *our appreciation of the world*. The mediaeval synthesis took it that the world reflected a divine order. Some continued this affirmative line, even if the order of Nature itself was seen differently. As the poet Pope wrote as the epitaph for Newton's tomb, 'Nature and Nature's laws lay hid in night; God said, Let Newton be, and all was light'. However, others felt that with the loss of the traditional order of the Universe, all sense of order – whether natural or social – was crumbling (see, for instance, Toulmin 1990, pp. 62–69). The changing appreciation of the world is exemplified by the cultural impact of the earthquake that destroyed Lisbon in 1755. The French philosopher Voltaire wrote a *Poème sur le désastre de Lisbonne* (1756). In Voltaire's book *Candide ou l'Optimisme* (1759) the philosopher Pangloss keeps defending the position that this world is the best of all possible worlds. The more Pangloss argues his case, the less convincing it becomes. Whereas the mediaeval synthesis affirmed the world as God's good creation, the present perception allows for meaninglessness and ambivalence among the spectrum of valuations of the world.

*Changes in our understanding of religion* may be related to all three developments mentioned above. Some have attempted to adapt contemporary theology to substantial changes in our view of the world. For instance, creation is no longer understood as a once and for all event, but rather as a continuous process. Such forms of adaptation to contemporary insights are to some extent necessary, but they have their problems. We are evolutionarily adapted to think in terms closely connected with common sense experience: 'the Sun rises' rather than 'the Earth turns'. Similarly, we are prone to imagining concepts like 'heaven' as 'above', even though it is hard in a time of world-wide flying to maintain that there is a throne on the clouds. It is not easy to free ourselves from the categories of thinking which were fruitful in dealing with the meso-level of reality that was relevant to survival in the evolutionary development of the human species. Hence, many people experience changes in our way of conceiving of reality as an unnerving loss, even though they may agree on the need for new images and concepts.

Theology has also responded to the emphasis on the human role in knowledge, for example by withdrawing to 'feeling' (Schleiermacher), by taking up Kant's transcendental argumentations about the conditions for the possibility of knowledge or of ethics, by turning to the subjective and personal (e.g. Martin Buber's 'I-thou' in contrast with 'I-it'), or by focussing on religious language and tradition.

Changes in appreciation of the world have affected theology as well. This is most explicit in those theologians who moved from an understanding of God in metaphysical terms, say God as the Ground of Being, to an understanding of God as being on the side of the victims or the poor. The 'Death of God' discussion of the 1960s reflects the stronger emphasis on human autonomy in creating knowledge (the previous point) and in responsibility, as well as a strong sense of the reality of horror, of injustice in the world.[4]

Thus, various authors writing on science-and-religion may easily talk past each other, even though they seem to address the same issues. Underlying presuppositions shape the way the dialogue is presented. We will discuss examples of the contemporary scene in two clusters. Are physics and theology engaged in a common quest for understanding? If so, how is the religious 'object' envisaged: as a God beyond reality, as a mystery at the heart of reality, or as meaningfulness (3)? And if religion is not understood as knowledge about some ultimate reality, how is religion to be seen in the context of the evolution of the human species (4)?

## 3 Three 'gods' in common cognitive projects

The cognitive relevance of science for religion may be clustered around three ways of thinking about ultimacy: God, mystery, and meaning, correlating more or less with the distinctions made above between changes in our knowledge, in our view of knowledge and in our appreciation of the world (2.2). How might one think about God in relation to the Universe (3.1)? Is mystery a persistent ground for religious wonder (3.2)? Is there ultimate meaning to human existence in the Universe (3.3)?

### 3.1 God

Empirical science arose, according to A. N. Whitehead (1926), when God was conceived of as endowed with 'the personal energy of Jehovah and with the rationality of a Greek philosopher'. The properties of the world could not have been deduced by thought alone (the Greek strand), but neither could they be taken to be purely whimsical, without regularities, totally dependent upon the mood of some deity. If one adopts this view of the rise of modern science, science and belief in God were allies rather than enemies. One could question this view of the history of science by pointing to other factors, such as the development of technology. However, the themes of *contingency* (3.1.1) and of *rationality* (3.1.2) are still central to discussions relating science and theology. After focussing on these two correlates of divine will and divine reason, we shall discuss 'design' as a more qualitative notion in this context, which may be related to specific intentions of the supreme being, such as its love for humans and its longing for a free response (3.1.3). In conclusion (3.1.4) various ways of understanding 'God' will be considered, such as a Platonist view, with emphasis on rationality, and a deist view, with emphasis on initial contingencies.

### 3.1.1 *Contingency: room for divine action?*

> . . . , any contemporary discussion regarding theology and science should first focus on the question of what modern science, especially modern physics, can say about the contingency of the Universe as a whole and of every part in it.
>
> (W. Pannenberg 1988, p. 9)

An event, property, or state of affairs is contingent if it is possible but not necessary; if it could have been different. The theologians Pannenberg and Torrance have correlated the contingency of the world with divine freedom. God might have chosen differently. God could have created no world at all, or a world with different ingredients (laws and initial conditions). In extremis one might hold the view that God would not be bound in any way; God could have created something logically contradictory or wicked. Most theologians avoid such consequences by including goodness and rationality in their concepts of God. If one includes relationality in one's concept of God, as process theologians do, God could not lack a creation either. Such a view would have no need for a contingency of the existence of reality.

Various *varieties of contingency* might be considered (e.g. Russell 1988). The contingency of existence refers to the question: why is there anything at all, rather than nothing? Initial contingencies may be presented in various ways. Why did the Universe start with the properties it has? Why did the Universe start at that moment (if seen from eternity), and not, say, infinitely long ago or at some other moment of time? And could the laws of Nature, or the constants involved in them, have been different? Aside from questions about the Universe as a whole, is there some form of 'local' contingency in the processes governed by the laws? The contribution by John Barrow in this volume can be read as an analysis of such contingencies in the context of contemporary physical theories, for example in his discussion of broken symmetries.

It should be clear that, again, not merely the answers are to be debated, but *the ways in which the questions are posed* as well. For example, one may ask whether the Universe had a beginning a finite time ago (the Big Bang theory) or not. One may then consider problems of the kind: 'What was God doing before he created the world?' Rather than answering this question, the theologian Augustine of Hippo (*c*. 400) replied that the question is wrongly posed, as time is created together with material reality, rather than being an uncreated universal background. A similar move has been made in contemporary cosmology, when the Newtonian concept of time and space as divine attributes was replaced by the general relativistic understanding of space–time as a physical entity. This move is even more explicit in quantum cosmology, where time is not a concept that is universally applicable (e.g. Isham 1991, 1993; Drees 1991, 1993). Thus, the alternatives are no longer an infinitely old Universe or a sudden beginning. Rather, it may be that 'time' loses its status as a universally applicable ontological notion. Similar shifts may occur for other types of contingency: rather than answering the question, one might reflect upon the question and reformulate it in a more appropriate way.

Contingency and necessity seem to me to be relative terms; something is contingent or necessary in the context of a theory. Science assumes a principle of sufficient reason, the heuristic principle that one should always seek reasons. However, this is a methodological principle rather than a metaphysical principle, which would claim that there always *must* be reasons. Absolute necessity is beyond science. And absolute contingency is also beyond science. Something which is contingent in the context of current theories may be shown to be unavoidable given certain circumstances.

There seems to be *an emotional resistance against contingency*, as it seems to make the course of life uncertain, a consequence of random events rather than of intentional decisions. 'Conspiracy theories' signal the emotional resistance against interpreting processes as contingent. If bad luck causes the loss of a dear one, some may attempt, in my opinion misguidedly, to comfort those that grieve by suggesting that the loss might be according to a higher plan. Thus, one might be religiously motivated in opting for a hidden variables view in quantum theory, according to which all, apparently random, correlations between events are traced back to a higher coherence.

The discussion about contingency is similar, though more ontological in language, to the discussion about complete theories and the reach of science (which is more epistemological in language, though not exclusively so, e.g. Barrow 1988, 1991 and in this volume). It seems closely connected with ideas about divine creation and interference.

The discussion may focus on global issues, such as 'the beginning' of the whole of reality. However, it may also be directed towards local contingency, concerning a part of reality. One might consider quantum theories and chaos theories as referring to local contingencies in reality. However, these theories can be interpreted deterministically as well. For example, Many Worlds Interpretations of quantum theories, which take it that all possibilities have equal ontological status, imply that there is no contingency of this kind, though the theory itself may still be considered contingent.

The interest in 'local' contingencies may be triggered by the quest to find an acceptable way of thinking about divine action in the world. Another interest might be to find a place for human free will. However, 'free will' should not be confused with randomness and unpredictability; it is more like 'self-determination'. Thus, philosophers have also argued that free will is compatible with full determinism (Dennett 1984). Some determinism is necessary for a meaningful understanding of free will, for otherwise it would be impossible to make plans and carry them out.

The reverse side of the emphasis on contingency may be exemplified through a reference to the polemical book by the British physical chemist Peter W. Atkins, *The Creation*. As he sees it, science has traced complex entities back to simpler entities, and these to the vacuum, which may have arisen by chance. He develops the view that

> the only way of explaining creation is to show that the creator had absolutely no job at all to do, and so might as well not have existed. We can track down the infinitely lazy creator, the creator totally free of any labour of creation, by resolving apparent complexities into simplicities, and I hope to find a way of expressing, at the end of the journey, how a non-existent creator can be allowed to evaporate into nothing and to disappear from the scene.
>
> (Atkins 1981, p. 17)

Removing contingencies by offering encompassing explanations would do away with tasks for a deity. One might nonetheless introduce an inactive deity, or even a deity which created it all by hand, but that would be superfluous, against the spirit of science (Occam's razor).

### 3.1.2 *Order and intelligibility: divine rationality?*

Whereas emphasis on the role of experiments in science seems to align with the interest in contingency, one could suggest that the role of mathematics in science aligns more with stress on rationality. The mathematical nature of the order of the Universe may suggest religious themes, especially to theoretical physicists and cosmologists. Sir James Jeans referred to God as a pure mathematician, and Paul Davies titled his recent book *The Mind of God: The Scientific Basis for a Rational Universe* (1992). And the Polish cosmologist and priest Michal Heller wrote:

> Mathematical structures that reflect the structure of the world are indeed able to provide a kind of ultimate understanding. In a theological perspective the ultimate rationality is that of God. The fact that it is a mathematical type of rationality is not a new factor in theology. The metaphor of 'God thinking the Universe' is well rooted in the history of theology.
>
> (Heller 1990, p. 207)

One might consider the intelligibility of the Universe, our capacity to grasp the regularities of the Universe, to be the amazing aspect which invites further reflection of a religious kind. Thus, the Canadian philosopher Hugo Meynell argued that the emphasis should be 'on the explanation of the intelligibility of the world, rather than on accounting for the gaps in that intelligibility' (Meynell 1987, p. 253). It is a kind of second order argument. Different cosmological programs suggest very different ontologies, but they have in common that they provide an intelligible Universe. What is it about the Universe and about us that makes it possible to think in an orderly way about it; in other words, makes it possible to frame adequate mathematical theories? Meynell defends the view that 'there is something analogous to human intelligence in the constitution of the world' (1982, p. 68). Paul Davies addresses similar questions in his recent book (Davies 1992) and in his contribution to this volume. He holds the view that the capacity to do the relevant mathematics is not evolutionarily explicable. I will use the opportunity to suggest some of the problems that the position of Davies might have to face.

(1) Is it true that the world is intelligible? Who understands quantum theory? Many are able to work with the formalism, but 'understanding' is an ambiguous term, as the following anecdote from an Oxford exam of about a century ago, quoted by Barrow (1988, p. 193), illustrates.

> Examiner: What is Electricity?
> Candidate: Oh, Sir, I'm sure I have learn't what it is – I'm sure I *did* know – but I've forgotten.
> Examiner: How very unfortunate. Only two persons have ever known what electricity is, the Author of Nature and yourself. Now one of them has forgotten.

Almost everybody knows that the light switch affects the light. Many believe they understand it. But those that continue questioning may well raise difficulties. Electricity is manipulable and well describable by Maxwell's laws. However, is manipulation and calculation the whole of understanding?

(2) The fact that mathematics is so effective in theoretical physics may be like the amazing fluency of natives in their own language. Physicists have been trained in using mathematics, so why should one be surprised if they use it and find it effective for the problems they deal with? It might well be a property of their approach or of the problems they select. However, such a rebuttal seems too easy. Mathematical science has an effectivity which is also recognized by those not trained in mathematics.

(3) An evolutionary explanation assumes that the capacity for knowing may have been advantageous to those who had it. It seems obviously advantageous to be able to anticipate the trajectory of a moving object, say a falling apple, a stone or a spear. However, what could have been the advantage for early hominids of being able to work out explicitly a mathematical theory of these phenomena? We seem to have been served more than we ordered. The ability to do abstract mathematics seems evolutionarily inexplicable – a product of design rather than of evolution?

The case for an evolutionary explanation might be defended by considering the similar problem for reading and writing. Whereas the ability to use speech seems to have co-evolved with certain structures in the brain, this cannot have been the case for reading and writing, which are fairly recent cultural innovations. In this case it seems likely that certain already existing structures in the brain were used for the new purposes of reading and writing, perhaps structures that previously served in interpreting tracks of animals (and were therefore evolutionarily selected). An inexplicable preadaptation of the brain to reading is an unnecessary assumption. Rather, the evolutionarily explicable plasticity of the brain allowed for these new forms of behaviour. Similarly, we may be able to use the ten fingers to play the piano, although they evolved due to other evolutionary pressures. And similarly, one might argue, there could be a potential for abstract mathematical reasoning, though the corresponding brain structures were developed for other purposes due to other evolutionary pressures.

(4) The apparent intelligibility and lawfulness of the Universe may also be a consequence of selective observation. One could imagine a chaotic Universe, with some temporal eddies of regularity and intelligibility. Complex organisms, such as human beings, might exist only in such eddies, and hence observe regularity and infer intelligibility. An analogy may be taken from economics. One might think that intelligible economic development correlates with a strong central planning agency. However, a free market economy may, in principle, also lead to an overall intelligible development even though there is no overall planning. One could interpret this by formulating 'freedom' as the ovcrall principlc, but onc could also say that thcre is no principle. This is the line of thought behind J. A. Wheeler's quest for 'law without law'. The trend in theoretical physics may be towards complete intelligibility, a unified Theory of Everything. 'But it may be that our undoing of the catalogue of Nature's laws will take us down a different road, which will lead us to the recognition that there is no such ultimate Theory of Everything: no law at all' (Barrow 1988, p. 297).

### 3.1.3 *Design and the anthropic coincidences*

Even if contingency, intelligibility or unity would allow an interpretation in relation to God as creator, they seem to lack specific intention. The specific order of the Universe might, however, be interpreted theistically as evidence that God is not interested in creating any odd universe, but a universe in which beings of a certain kind can exist.

The most recent variant of such thinking has been debated in relation to the anthropic principles (e.g. Barrow and Tipler 1986, Davies 1982, Drees 1990, pp. 78–89; see also Barrow and 't Hooft in this volume). The Universe has certain properties, for example three spatial dimensions, a certain size and age, a specific strength of the gravitational force compared with the electromagnetic force. One might consider other universes with other properties. It turns out that most of such imagined universes would not allow for the evolution of complex organisms based on carbon chemistry. Thus, we seem to be extremely lucky that the Universe actually has the properties it has.

The discussion needs some distinctions. First, one should distinguish between the *anthropic coincidences*, as I prefer to name correlations between properties of the Universe and our existence, and the *anthropic principles*, which may be seen as various attempts to interpret the coincidences. Among these are the Weak Anthropic Principle, which points out that what we see is biased in favour of those parts of the Universe where we can exist, thus emphasizing that the coincidences may have to do with selective observation. There are also Strong Anthropic Principles, which suggest that these properties are in some sense necessary for any possible universe, whether observable or not.

Within this latter category one might consider as a specific case a Theistic Anthropic Principle: these properties are a consequence of design. An attractive feature of the design argument is that it suggests something about God's intentions, of which humans are a part. Perhaps God created our kind of world because God wanted a world in which living, conscious and sentient beings could live and relate to God in a freely loving way. Freedom of response might have been such a highly valued good for the creator, that the creator was willing to sacrifice full control of the processes. Along such lines some have suggested that belief in a powerful and loving God might be compatible with the evil and suffering present in the world (a free will and free process defence).

I have doubts regarding the anthropic arguments and their religious use, which I will present below.

The *Weak Anthropic Principle* (WAP) states that what we see must be compatible with our existence. We see a Universe with planets, as our existence depends on planets. We see a Universe which existed for billions of years, because it took billions of years to develop beings which are capable of thinking about the age of the Universe. The WAP has the nature of a selection rule: our observations are biased in favour of situations where we can exist. It is as if we attempt to find out how often railway crossings are closed. If we observe them only from trains, we will find that crossings are always closed. In cosmology, we can observe the Universe only from a spatially and temporally very restricted set of points of view. Thus seen, the WAP has no metaphysical significance but is a reminder of the biased nature of our observations.

I think that the WAP is correct but devoid of relevance. If we know that life depends on liquid water and we observe the existence of life, we may conclude that our environment must contain water, and thus must have a temperature within a specific range. This is the common use of evidence: we observe life and conclude that something else which goes together with life is also the case. This does not explain either our existence or the existence of water. The explanation of an event or of certain conditions is in general something different from the explanation one offers when asked 'How do you know?' From the existence of this chapter you may infer the existence of its author, but the paper does not explain my existence. Retrograde reasoning justifies beliefs, but it does not explain why the situation was the way it was.

The WAP might be combined with the idea that *all possible worlds are actual*. The latter has been defended, for example in analogy with a Many Worlds Interpretation of quantum theory or in the context of specific cosmological theories which allow for many different regions (larger than our observable universe) with different properties. The existence of our Universe with its anthropic coincidences is then explained on the basis of the assumed actuality of all possibilities. The explanation is not so much due to the principle of selective observation (WAP), but to the metaphysical view that all possibilities are actual (plenitude), which 'explains' the existence of quite a lot.

The *Strong Anthropic Principle* (SAP) has been stated thus: 'The Universe must have those properties which allow life to develop within it at some stage in its history' (Barrow and Tipler 1986, p. 21). This is not a statement about what we actually observe, but about the class of possible universes. It leads to an explanation of properties of the Universe in terms of purpose: a property that is necessary for life is necessary for the Universe. Such teleological explanations have a long history, but they are not widely accepted in contemporary science. SAP arguments have some disadvantages. (1) Properties of other possible universes are unobservable and untestable. Thus, strong anthropic reasoning cannot rely on testable consequences about the class of possible universes, but must base its appeal on the coherence of the view which it supports. (2) Is it helpful to explain the properties of the Universe by reference to life or consciousness, which in all its richness (and possible other forms) is not fully understood? (3) SAP explanations are vulnerable to future developments. Successive theories have, generally speaking, fewer and fewer parameters. If that trend continued, the set of possible universes (without invoking SAP reasoning) might be very small, perhaps even containing only one possible universe, ours.

If applied at a smaller scale, as in 'planets must have the properties which allow for the development of life in some stage of their history', a Strong Anthropic Principle is surely false. However, the example shows the teleological nature of SAP: everything must have a purpose. Hence, the Moon must be populated, as the ancient philosopher Plutarch argued.

*A Theistic Anthropic Principle?* John Polkinghorne, a professor of theoretical physics who became an Anglican priest, calls the ideas about other worlds 'metaphysical speculation'. 'A possible explanation of equal intellectual respectability – and to my mind greater economy and elegance – would be that this one world is the way it is because it is the creation of the will of a Creator who purposes that it should be so' (Polkinghorne 1986, p. 80).

Such an apologetic strategy does not work. (1) The argument assumes that the anthropic coincidences are here to stay. However, some (or even all) of these features, which apparently point to design, might find more traditional scientific explanations in future theories. That has happened for the traditional design arguments based on intracosmic adaptedness. The inflationary scenario in cosmology has already led to some erosion of anthropic coincidences. (2) The idea that the assumption of a single creator is more economical than the argument of a plurality of worlds (with WAP selection) is a misapplication of the 'economy' rule (Occam's razor). It is simpler to accept a theory, say about the formation of planetary systems, in its predictions beyond the observable domain, than to draw a line between what is currently observable and the rest. The issue of simplicity and economy is not so much one about the number of entities predicted by a theory but one about the structure of the theory. Some cosmological theories are more simple if one allows for the existence of many worlds than if these were to be excluded. (3) The argument for design assumes that certain features of the Universe are improbable if not for design. However, the probability or improbability depends on the features considered and on the class of alternative universes considered. The basis of the argument is not as objective as it seems, as has been pointed out by Ovenden (1987, pp. 105f.).

### 3.1.4 *What kind of God?*

> 'The most miraculous thing is happening. The physicists are getting down to the nitty-gritty, they've really just about pared things down to the ultimate details, and the last thing they ever expected to happen is happening. God is showing through.'
>
> 'Mr Kohler, What kind of God is showing through, exactly?'
>
> (Dialogue between a computer freak and a professor at a divinity school, in *Roger's Version* by John Updike (New York 1989, p. 9)

As in the quoted dialogue, it is relevant to ask 'what kind of God' is appearing in the context of such cognitive dialogues between science and religion.

'God' may be thought of as the cosmic watchmaker, the engineer who constructed the initial state and lit the fuse. Carl Sagan wrote in his preface to Stephen Hawking's *A Brief History of Time* that the consequence of Hawking's theory is that there is no absolute beginning of reality, and therefore no need for a creator (Sagan 1988, p. x). A similar *deist* notion of God seems the aim of arguments that purport to show that there was an absolute beginning, inexplicable within the Universe, and hence a cause beyond the Universe.

When Hawking concluded his book by linking knowledge of the ultimate theory to knowledge of 'the mind of God', he used a more *Platonist* image of God. Other examples of this kind of thinking among theoretical physicists are Roger Penrose's defence, in his *The*

*Emperor's New Mind*, of the reality of a timeless realm of mathematical truths, and Paul Davies' discussion, in his *The Mind of God*, of the relation between mathematics and reality (see also the essays by 't Hooft and Davies in this volume).

A *theist* view has God both as the highest (transcendent, timeless) being, a Platonist element, and as the original creator, a deist element. Besides, God is understood to be active in time, either in human history or in the whole course of evolution (*creatio continua*). This position is not an easy one to maintain, since these aspects of God are hard to combine with each other.

An alternative may be a *pantheist* view, which assumes an ontological identity of God and world, rather than the more dualist conceptions mentioned before. We will return to such views in the next section, on mystery.

### 3.2 Mystery: a common awareness of not-knowing?

Robert Jastrow concluded his *God and the Astronomers* with the following image.

> For the scientist who has lived by his faith in the power of reason, the story ends like a bad dream. He has scaled the mountains of ignorance; he is about to conquer the highest peak; as he pulls himself over the final rock, he is greeted by a band of theologians who have been sitting there for centuries.
>
> (Jastrow 1980, p. 125)

The essence of modern cosmology is, according to Jastrow, that the Universe 'began at a certain moment of time, and under circumstances that seem to make it impossible – not just now, but ever – to find out what force or forces brought the world into being at that moment' (1980, p. 12). Theology always lived with the awareness of its inability to express what God is. This section will not deal with the cosmological issue; Jastrow's specific example is in need of modification due to the development of quantum cosmology (Isham 1991, 1993). The issue here is the emphasis on the limits of human knowledge. Is there a common meeting ground for religion and science in not-knowing?

One of the Ten Commandments in the Jewish and Christian heritage is the prohibition against worshipping idols, a practice which is considered religiously and socially destructive. The Greek heritage developed a more metaphysical and epistemological critique of anthropomorphic concepts of God. As the origin of knowledge and existence God is beyond knowledge. Later systematic thought in Christianity distinguished between two ways of thinking about God's attributes. The first way is one of extrapolation and affirmation. We know to a certain extent what 'power', 'presence', and 'wisdom' mean. God is then thought of as omnipotent, omnipresent and omniscient. The other approach is labelled the *via negativa*: we deny features of reality in reflecting upon God. Stating that God is atemporal is not a positive statement about God's nature, but a denial of temporality. 'God is infinite' is not a cognitive statement, as if one claimed to know what 'infinite' meant, but a denial of creaturely finitude. We should respect the 'is not' character of metaphors, especially in religion, as they save us from absolutizing images and falling into idolatry (McFague 1982). That would result in a loss of sensitivity for the symbolic nature of religious language. Thus, recognition of God as the unknowable, as a mystery at the heart of religion, seems well rooted in religious thought.

Are there any reasons within science to assume that there is a mystery about which we cannot speak scientifically? In his *Cosmic Understanding* (1986) Milton K. Munitz gave a philosophical analysis of scientific cosmology. The Universe as it is known, as an intelligible unit, is a product which results from the application of a conceptual scheme. This should not be misunderstood, as if reality owes its existence to concepts. It is an epistemological point: all theories about the Universe are constructs which use human concepts. Conceptual boundedness is inescapable if one wants to achieve intelligibility. One aims at transcending the conceptual limitations of a theory by entering another conceptual scheme, which has its own boundaries. A similar point has been made by Harrison in his *Masks of the Universe* (1985). Each understanding of the Universe is a mask which is held in front of the real, but in itself unknowable, Universe.

One could stop here. Is there 'a dimension of reality "beyond" any account of the known Universe (or any of its contents), of which we can have a mode of awareness that is not hemmed in by the constraints and ever-present horizons of cosmological knowledge?' (Munitz 1986, p. 229). The epistemological endlessness of the cosmological search might be due to such an ontological 'ultimate boundlessness or indeterminability at the very heart of reality' (Munitz 1986 p. 229). This reality would not be conceptually bounded the way the Universe is, nor would it be bound by anything beyond itself. 'Boundless Existence' is not the name of an object or entity.

> We shall be driven, consequently, and at the end, to silence, although the 'talk' on the way, if at all helpful, will have its value in making the silence a more pregnant one, and indeed the occasion for having an overridingly important type of human experience.
>
> (Munitz 1986, p. 231)

Munitz attempts to point to something which epistemologically transcends all our knowledge – something which is, however, at the heart of reality, hence ontologically immanent.

In the context of quantum physics, Bernard d'Espagnat has argued that 'we have to reckon with two realities. More precisely, present-day physics calls for a clear-cut distinction between two notions both designated in the past by the word 'reality', independent reality which is distant, 'veiled', and empirical reality, the totality of phenomena (d'Espagnat 1989, p. 7). It is this latter reality that we understand better with each day that passes. Positivist thinkers, both from philosophy and in the Copenhagen tradition of quantum physics, have attempted to rule out independent reality as meaningless. Materialists and realists tend to subsume the notion of empirical reality under that of independent reality. However, 'in our time science itself has provided us with pressing reasons for accepting the (philosophical) duality of Being and of phenomena' (d'Espagnat 1989, p. 7). These reasons are taken from quantum physics, which is not merely about the nature of reality, but about the possibilities of knowing about reality.

Does science suggest the existence of an unknowable or veiled reality? This question is strongly linked with discussions regarding scientific realism with which various chapters in this volume deal. Besides, one should not merely consider what we do not know (which might be knowable or unknowable), but also what we know negatively. Ideas previously held to be true, or probable, have been shown to be wrong, or at least probably wrong. Science may not be able to provide a grand view of reality, including the mysterious, but scientific research may serve well to criticize ideas about that reality. This could be liberating, as it creates room for new ideas.

If the 'veiled reality' (d'Espagnat) or 'boundless existence' (Munitz) is taken seriously, one still may wonder what its significance for religion can be. Is it merely an expression of our cognitive limitations, and the recognition of unknowable or partly knowable aspects of reality? Is that sufficient to inspire religious awe? Or does religious mystery presuppose certain qualities about that mysterious reality, which might justify associations with love, trust or beauty? It is not clear to me how such a transition of categories is to be made in the context of these approaches.

Besides, there may well be a strong hesitancy in theological circles to make too much out of these apparent mysterious aspects. The methodological catch word is the 'God-of-the-gaps'. It has happened that gaps in a scientific account, e.g. an account of the evolution of the human out of earlier mammals, or of complex physical phenomena, were seen as possible loci of special divine intervention. Such an approach disregards the coherence of the scientific account, and may result in a religious position which is always on the retreat as science successively fills such gaps. Are there gaps which do not erode? Are these mysteries of Munitz and d'Espagnat persistent? Or will there be a future theory beyond quantum physics which lends itself much better to an interpretation of reality in a single scheme of objective reality, without the distinction between veiled and empirical reality? One might connect this with the earlier discussion about contingency: is there any contingency that will be inaccessible to science? Are the laws of nature possible candidates, or might they be necessary? And what about the existence of something rather than nothing?

### 3.3 Meaning: human existence in the Universe

Neither 'God' nor 'mystery' is the concept central to contemporary religious thought. 'Meaning' seems to have become a replacement which is more acceptable to many in contemporary secular Western culture because it is less suggestive of anything supernatural.

*Order out of Chaos* (1984) by Ilya Prigogine and Isabelle Stengers has as its title in the original French edition *La Nouvelle Alliance: Métamorphose de la Science* (1979). The claim is that there is a new alliance between humans and Nature, due to changes in science. The classical (Newtonian) physical sciences used to think of reality in a way in which human existence was itself a marginal side product of the evolutionary process of mutation and selection. But developments in science are believed to have paved the way for a new view of the place of humanity in natural reality. Unlike the covenant of Moses at Mount Sinai, this one is an alliance between humanity and physical reality. Humans are no longer strangers in a mechanistic world. The central tenet of Prigogine and Stengers is that 'science is rediscovering time' (1984, p. xxix). Their case is based upon the development of thermodynamics for systems which are far from equilibrium, which, under the right circumstances, develop into new ordered states.

When Nobel Prize winner Prigogine and his co-author Stengers claimed a 'new covenant', they were responding to Nobel Prize winner Jacques Monod whose influential book *Chance and Necessity* ended with the following sobering (or liberating?) thought:

> The ancient covenant is in pieces; man at last knows that he is alone in the unfeeling immensity of the universe, out of which he emerged only by chance. Neither his destiny nor his duty have been written down. The kingdom above or the darkness below: it is for him to choose.

The kingdom above is the kingdom of knowledge, 'within man, where progressively freed both from material constraints and from the misleading servitudes of animism, man could at least live authentically' (Monod 1971, p. 167). The 'darkness below' is the variety of animisms, including utopian ideologies such as historical materialism. The ethics of knowledge is based on an ethical choice, an axiom which humans impose on themselves. It 'thereby differs from animist ethics, which all claim to be based on the "knowledge" of immanent, religious or "natural" laws which are supposed to impose themselves on man' (Monod

1971, p. 164). Animisms fail to make the proper distinction between judgements of value and those of knowledge.

> It is perfectly true that science attacks values. Not directly, since science is no judge of them and *must* ignore them; but it subverts every one of the mythical or philosophical ontogenies upon which the animist tradition, from the Australian aborigines to the dialectical materialists, has based morality: values, duties, prohibitions.
>
> If he accepts this message in its full significance, man must at last wake out of his millenary dream and discover his total solitude, his fundamental isolation. He must realize that, like a gypsy, he lives on the boundary of an alien world; a world which is deaf to his music, and as indifferent to his hopes as it is to his suffering or his crimes.
>
> (Monod 1971, p. 160)

Whereas Prigogine and Stengers may be seen to argue for close ties between humanity and the cosmic processes, Monod describes humanity as a cosmic oddity, arisen by accident. Meaning is not found in that process, which is described by science, but rather in a more existentialist mood in the human choice for objectivity. Objectivity as the ethical axiom cannot itself be based upon some scientific objective basis. It is this ethical axiom which bars science from becoming a basis for further values.

The merits of various proposals about science and meaning, also in the context of other sciences, deserve, of course, more detailed discussion than can be given here. I want to conclude by suggesting a position which is intermediate between meaninglessness and meaningfulness. The astronomer Hubert Reeves refers to the myth of Prometheus when he discusses the development of nuclear weapons. An unprecedented capability for destruction is based on an impressive amount of knowledge. We are placed in a border zone, between good and evil. The development of complexity, life, consciousness and intelligence in the course of cosmic evolution is ambivalent. Meaning is something we may create, rather than detect.

## 4 Naturalist challenges for religion

So far we have concentrated on different views of the 'object of religion', God, mystery and meaning, correlating these with changes in knowledge, views of knowledge, and in appreciation of the Universe. Now we will take our starting point from the scientific side. It will be argued that science poses serious challenges to religion; not just to religion which takes science to be relevant, but also to religion which seeks mutual neutrality through separation (see above, 2.1). We will say a few words about the nature of science, before turning to scientific approaches to religion.

### 4.1 The nature of science

> Mountain peaks do not float unsupported; they do not even just rest upon the earth. They *are* the earth in one of its manifest operations.
> (John Dewey, *Art as Experience*, 1934, p. 3)

The motto of this section intends to capture an outlook which is shared by most people involved in science. Mountain peaks do not float unsupported, nor do they rest upon the Earth. Rather, they are manifestations of the Earth. Similarly, expected and unexpected phenomena are assumed to be part of the same physical reality. When superconductivity at higher temperatures was discovered, no one doubted that that phenomenon would be, if confirmed, a manifestation of the possibilities of matter, even if it was not yet understood how it fitted into existing theories. Such an assumption, taking all phenomena to be manifestations of a single reality guided by the same basic laws, correlates with the methodological rule that one should search for naturalist explanations similar to those that were adequate for simpler phenomena. Thus, explanation of biological phenomena in chemical terms will be attempted, even though some biological concepts go beyond the vocabulary of chemistry. Ontological monism without reductionism of all sciences to physics is one of the contemporary positions presented by Dieks in his essay in this volume. Even if one agrees that complete reduction to physics is impossible, the attempt may be heuristically useful. Since the basic rules of physics and chemistry do without 'purpose' or 'goal directedness', one is led to the idea that biological evolution should be explained without reference to an overall purpose, even though individual organisms do have purposes, if described at a level which allows for such a concept.

In this context, two methodological rules are often invoked, though it is difficult to express them both in a precise form and with general applicability. The first rule, 'Occam's razor', named after the fourteenth-century theologian William of Occam, demands that one should not introduce additional entities or principles without good reason. What counts as good reason needs, of course, further discussion in any specific situation. Using this rule, adding a religious explanation when one could do without one, is against the spirit of science.

The other methodological rule, requiring the avoidance of *ad hoc* assumptions, works against religious superstructures as well. This can be illustrated by the way P. H. Gosse harmonized Darwinian evolution with his belief in a fairly recent creation of all species. In his book *Omphalos* (navel) (1857) he argued that God had created Adam with a navel, suggesting a mother where there had been none. The trees in the Garden of Eden had been created with a complete set of annual rings, suggesting growing seasons though there had been none. The fossils suggest an evolutionary history, but one can not refute the possibility that God created them in the appropriate strata. This view is logically consistent, but contrived, *ad hoc*. All extragalactic and most intragalactic astronomy as well as palaeontology would become a study of God's magic rather than a study of stars, galaxies, and of species that once lived on Earth. It may be hard to specify precisely when a move is to be considered *ad hoc*, but that does not take away the intuition that certain moves are to be avoided.

We will now turn to the understanding of the phenomenon 'religion' under the monist assumption. However, it should be noted that not all working scientists nor all philosophers agree on the assumed ontological monism (see, for instance, Feyerabend, Chapter 6, in this volume).

### 4.2 Naturalist views of religion

Reflections on God, meaning and mystery assume that religion embodies significant cognitive insights. Some serious thinkers, both scientists and theologians, see religion not as a metaphysical supplement to scientific explanations, but as dealing with other aspects of human existence. As will be argued here, such a view of religion does not imply that there is no need for a dialogue with the natural sciences. Religion may be far removed from physics, but an existentialist, ethical, or anthropological view of religion gets one into a discussion with naturalist approaches in biology and, beyond biology, in the human sciences.

Humans, with their epistemological capacities and their habit of making moral judgements, evolved from other hominids in correlation with the evolution of culture, e.g. the use of language and the ability to perceive and communicate various states of affairs. According to Alexander (1987), mechanisms of indirect reciprocity and social status within a group and the need for coherence, and mutual support in the competition with other groups of hominids, may have been important evolutionary pressures in favour of morality. The advantage of deceiving others, and the advantage of recognizing attempts at deception, may have been important factors in the evolution of the unconscious and of consciousness.

One might adopt a similar diachronic and functional approach to the origin of religion. Maybe religions arose in the evolutionary history of the human species as an essential ingredient for the emergence of distinctive characteristics of the human species, for instance by making possible large communities, inexplicable by relations of kinship alone. Burhoe (1981) argues for a co-evolution of human genetic and cultural information, with religion (ritual, narratives and more systematic systems of thought) imprinting cultural values into the human brain. As he sees it, religions have been essential to the evolution of the brain. Ritschl (1984, p. 34) discusses religion in the context of cultural pressures in a far more recent phase, as a way of dealing with stress when hunter-gatherers became sedentary. Whereas Burhoe, a Unitarian of Calvinist background, interprets evolutionary history in the context of an understanding of God as sovereign, powerful and selective, the Lutheran Theissen (1984) puts emphasis on tolerance for variation and grace (cf. Drees 1991a, pp. 92–100). Theissen seeks through a functional approach an ontological perspective on God or 'ultimate reality', understanding it as the realm to which successive adaptations adapt. Even if religion (ritual, myth and systematic reflection) arose as a consequence of certain evolutionary pressures, that does not exclude the possibility that there is a genuine transcendent referent of religious worship. However, in adding such a transcendent referent one is in danger of violating Occam's razor.

Besides the evolutionary perspective, one might also consider genetics, neurophysiology and other biochemical approaches to human nature. Is it possible to discuss human religion and morality independently of studies of such underlying structures? Reflections on genes and identity may well bear upon religiously relevant notions such as soul and body, spirit and flesh, self, immortality and resurrection, death, mystery (as the wisdom encoded in our genetic heritage is only secondarily accessible in language; Eaves 1991, p. 499), sin and grace (Eaves and Gross 1990).

Whether there is a role for physics apart from biology, is not easy to judge. The theologian Pannenberg (1985; see for a criticism Eaves 1989) prefers to stick to the work on open systems of Prigogine and others when discussing the phenomenon of life, whereas he turns to anthropologies of a philosophical nature when discussing the distinctively human, thus avoiding a confrontation with sociobiology and genetics. John Polkinghorne (1991) sees chaos theories as a first approximation to a physical theory which allows for human freedom and for divine action in the world. Appealing to contemporary physics in arguments regarding determinism, free will and self-determination is not easy (e.g. Earman 1986). Some problems arise due to the implicit leap from physical discourse to the discourse of the humanities. The intermediate levels of biological reality enrich the conceptuality, both with respect to persons and environments, in such a way that the discussion changes. On the one hand, using physics to avoid confrontation with geneticists and sociobiologists seems an unwarranted eclectic approach. On the other hand, is it possible to defend the concept of a free will without allowing for various possible outcomes, and hence without having an ontology which allows for various possibilities in connection with actuality? Or is a fully determinist view, without local contingencies, acceptable even if one wants to maintain notions like free will and responsibility, as Dennett (1984) defends?

As this section illustrates, even if one separates religion and science by emphasizing that religion has to do with personal aspects whereas science deals with non-personal aspects, one might need to reflect on science, especially where it deals with our view of human nature, including its social and emotional aspects.

## 5 Niches for religion

Various options for religion in relation to the natural sciences are available. One option might be to dismiss the scientific results. Such an approach, as has arisen in relation to evolutionary biology, seems to neglect the coherence of the sciences (see, for example, Kitcher 1982). Thus, the following will be restricted to approaches which intend to take science seriously. One option, not in all respects unlike the fundamentalist antagonism, is the dismissal of religion as being in direct conflict with science, or, at best, as having become totally superfluous. This line has been presented in an elegant, polemical way by P. W. Atkins (1981). However, before dismissing religion altogether, we may consider whether there are any other options available.

*Playing down the naturalist account* seems to be one possible strategy. Perhaps the case for science was overstated. If science consists of many theories which are successful in some restricted domain without justified pretensions regarding a greater unity, the assumed ontological monism might be discarded fairly easily. This option seems to characterize the essays of Hesse and Feyerabend in this volume.

If such a line is followed, science is still relevant to religion in two ways. (1) In those domains where science has successful theories, one should take good care to avoid contradicting them. Science may perhaps be unable to provide an encompassing theory, but there are many discoveries about the way the world is, or even more important, about the way the world is not. Clinging to refuted ideas, like a flat Earth, would inhibit the plausibility of the religious convictions espoused. (2) Science and religion are relevant to each other by contributing to a clearer view of the proper scope and limits of each enterprise. Thus, conflicts will be avoided by distinguishing more carefully between the businesses of science and of religion (e.g. Brümmer 1991, pp. 10ff.).

One might also opt for a niche *within the naturalist account*. It would be too easy, in my opinion, to claim divine intervention in order to explain phenomena, such as the origin of life, the extinction of the dinosaurs or the realm of quantum uncertainty. Such a God-of-the-gaps is in danger of having to retreat when science advances. Specific human feelings may seem to be outside the scope of the naturalist approach. However, confidence in the possibilities of regular explanations of such phenomena is well expressed by 't Hooft. The size and age of the Universe are to be expressed in such enormous numbers (counted in Planck times and Planck lengths) that this allows for 'feeling', 'free will', 'life' and something like awe and reverence for the immensity of the Universe ('t Hooft 1992, p. 231).

One might challenge the adequacy of naturalist approaches to religion. However, one might also accept those explanations, but add that such explanations do not bear on the possibility of a genuine religious response, as one could always exclaim 'Was that God's way of guiding me into faith?'[5] Such an overlay on top of the naturalist explanation is superfluous, and thus in danger of being shaved away by Occam's razor, even though there is no inconsistency. Once my children learn that gifts come from parents and grandparents, they might add that that is Santa Claus' way of providing for the gifts. However, they will soon start to look differently upon the whole happening, though it might remain a nice family event. The experience changes as a consequence of the newly acquired understanding.

There may be a relevant distinction between the description 'from outside', and one from within a specific perspective. The experience from within may be significant, even though another description 'from outside' is available. Without much hesitation we accept that the Earth rotates, while we experience the Sun setting. The experience is still there, but now tied to our specific perspective. Even if the 'thou' metaphor which we use in speaking to God would be 'metaphoric, something is lost when we attempt to translate the religious reality to the language of 'it', much as the joy of sex is not always enhanced by understanding the neurobiology of orgasm' (Eaves 1991, p. 502). Though some pioneering work has been done (e.g. Burhoe 1981, Theissen 1984), we still need a clearer view of the importance of religious life if one accepts functional views of religion and morality.

One might also attempt to find a niche *apart from the naturalist account*. The proper role of science in relation to religion could be in ethical considerations. Medical ethics is a growth business; the physicists, chemists and biologists have their issues. Among these are nuclear weapons, environmental pollution and genetic engineering, all of which have attracted public attention. Besides, one might reflect on the economical ethics involved: what moral considerations are involved in spending so much money on science? Which science would be justified? Science needs cultural support which values the search for knowledge, even if it is not profitable in a direct sense.

It may be questioned whether such ethical issues are within the province of religion. It is not only science which has emancipated itself from religion in modern times, but ethics and politics too. In Western, liberal societies various religious traditions co-exist under a large umbrella of allegiance to public laws and procedures and to a general set of human rights and human values. Religion has lost its grip on the public realm, but also on the private realm, where increasing numbers of people tend to base their choices on their own preferences or decisions rather than on allegiance to some religiously prescribed set of behavioural codes. The contribution of religion should, perhaps, be located at a deeper level, in the underlying existential attitude. For example, in discussing medical ethics one comes across views regarding human finitude and death. Are we willing to accept finitude? And how do 'life' and 'quality of life' count? Thus, from ethics one enters into a domain of existential questions.

It is not clear whether ethics is itself a safe niche for religion, as ethics too has been approached in a naturalist way, for example by Richard Alexander in *The Biology of Moral Systems* (Alexander 1987). However, even if all actual ethical systems are to be seen as strategies within specific cultures, they nonetheless may be taken to suggest absolute values as a kind of transcendental regulative idea beyond all cultural differences. Such absolute values are not available. This feature should keep us from fanaticism (Sutherland 1984). Whether we can say that these absolute values exist is strongly dependent upon the notion of 'existence'. It will not be like the existence of tables and chairs, nor that of electrons and magnetic monopoles. An analogy with mathematics might be more fruitful. Mathematics exists as far as it has been developed by humans, written down in textbooks and articles.

However, it does not seem as arbitrary as most human artefacts, such as literature. Mathematical theorems seem to have a truth which precedes their discovery.

The last option to be considered is a niche *above the naturalist account*. As the contribution by Paul Davies in this volume shows, one might ask various questions about the scientific descriptions as a whole. As Ernan McMullin wrote (1988, p. 74): 'The appeal is not to a "gap" in scientific explanation but to a different order of explanation that leaves scientific explanation intact, that explores the possibility for there being *any* kind of explanation'. This does not offer an argument for the existence of God within the context of science, as it is that context itself which is the focus of philosophical reflection. Introducing the notion 'God' in this context is not really an answer, an explanation for the Universe or for God. It points to an open place and keeps us aware of the questions.

McMullin and Davies agree in reflecting upon the whole of the scientific enterprise rather than upon apparent gaps within a scientific account. If a sense of wonder, reverence or the like is sufficient to qualify someone as religious, then both are religious. However, the normative and ritual aspects of religion are missing from the approach of Davies. McMullin maintains a richer notion of religion by taking seriously the religious tradition as it arose in Israel. As he formulates it, 'God is not to be seen *in* the universe, then, but *through* it' (McMullin 1988, p. 59). Mary Hesse, in this volume, aligns more with McMullin when she emphasizes actual human practices and traditions. She does not offer an argument for God for those taking a standpoint outside a specific tradition. Rather than looking for scientific contributions to a metaphysical theology of nature in the traditional sense, she holds it to be more fruitful to regard these reflections on science as 'debates about an appropriate *language* for theology, and a source of appropriate *models*' (Hesse 1981, p. 287).

Hesse and Davies address different questions, but neither of them addresses the agenda of the other in his or her own terms. This confirms the claim in the first paragraph of this contribution: discussions in science and religion are not merely discussions about answers, but also disagreements about the problems, and hence about the questions. The difference between Davies and Hesse also points to a problem with the various niches described above. One might take seriously the idea of some cosmic God beyond the naturalistic account, along the lines suggested by Davies, but would one ever light a candle to such a God? And if one takes the phenomenon of religion seriously as an important human practice, as does Hesse, one is in danger of ending with a religion without referent, a religion without God. How is one ever to combine such different strands and the transcendental regulative idea of absolute value?

## Notes

1. *U.S. News and World Report*, 23 December 1991, p. 59.
2. For Galileo's use of Scripture, see McMullin (1981); for a more socio-political emphasis, see de Santillana (1955); Copernicanism and exegesis are central to Langford (1966), e.g. pp. 65 ff.
3. Translated in S. Drake, *Discoveries and Opinions of Galileo* (New York: Doubleday, 1957); quotations have been taken from pp. 197 (twice), 186 and 187.
4. For instance, Surin (1986) rejects a theoretical theodicy. One should ask what we as God's creatures do to overcome evil and suffering rather than ask whether existence of an omnipotent and loving God and the amount of evil and suffering are compatible.
5. As the Dutch philosopher of religion Vincent Brümmer argued during the conference on which this book is based.

## Bibliography

Alexander, R. D. 1987. *The Biology of Moral Systems*. New York: Aldine de Gruyter.

Atkins, P. W. 1981. *The Creation*. Oxford: Freeman.

Barrow, J. D. 1988. *The World within the World*. Oxford: Clarendon.

—1991. *Theories of Everything*. Oxford: Clarendon.

Barrow, J. D. & Tipler, F. J. 1986. *The Anthropic Cosmological Principle*. Oxford: Clarendon Press.

Brooke, J. H. 1991. *Science and Religion: Some Historical Perspectives*. Cambridge University Press.

Brümmer, V. 1991. Introduction: A dialogue of language games. In *Interpreting the Universe as Creation*, ed. V. Brümmer. Kampen (NL): Kok Pharos.

Buckley, Michael. 1987. *Religion and the Rise of Modern Atheism*. New Haven: Yale University Press.
—1988. The Newtonian settlement and the origins of atheism. In *Physics, Philosophy and Theology: A Common Quest for Understanding*, eds. R. J. Russell, W. R. Stoeger, G. V. Coyne. Vatican Observatory; distributed by University of Notre Dame Press.
Burhoe, R. W. 1981. *Towards a Scientific Theology*. Belfast: Christian Journals Ltd.
Davies, P. C. W. 1982. *The Accidental Universe*. Cambridge University Press.
—1992. *The Mind of God*. New York: Simon and Schuster.
Dennett, D. C. 1984. *Elbow Room: Varieties of Free Will Worth Wanting*. Cambridge MA: MIT Press.
Dillenberger, John. 1960. *Protestant Thought and Natural Science: A Historical Interpretation*. Doubleday. (Reprinted 1988, University of Notre Dame Press.)
Drees, Willem B. 1990. *Beyond the Big Bang: Quantum Cosmologies and God*. La Salle: Open Court.
—1991. Potential tensions between cosmology and theology. In *Interpreting the Universe as Creation*, ed. V. Brümmer. Kampen: Kok Pharos.
—1991a. *Heelal, mens en God: Vragen en gedachten*. Kampen: Kok Pharos.
—1993. A case against temporal critical realism? Consequences of quantum cosmology for theology. In *Quantum Cosmology and the Laws of Nature*, eds. R. J. Russell, N. Murphy, C. J. Isham. Vatican Observatory & University of Notre Dame Press.
Earman, J. 1986. *A Primer on Determinism*. Dordrecht: Reidel.
Eaves, L. J. 1989. Spirit, method and content in science and religion. *Zygon* **24**: 185–215.
—1991. Adequacy or orthodoxy? Choosing sides at the frontier. *Zygon* **26**: 495–503.
Eaves, L. J. & Gross, L. M. 1990. Theological reflection on the cultural impact of human genetics. *Insights: The Magazine of the Chicago Center for Religion and Science* **2** (2 December): 15–18.
d'Espagnat, B. 1989. *Reality and the Physicist*. Cambridge University Press.
Harrison, E. 1985. *Masks of the Universe*. New York: Macmillan.
Hawking, S. W. 1988. *A Brief History of Time*. New York: Bantam Books.
Hefner, P. 1981. Is/Ought: A risky relationship between theology and science. In *The Sciences and Theology in the Twentieth Century*, ed. A. R. Peacocke. Stocksfield: Oriel Press.
Heller, M. 1990. The experience of limits: new physics and new theology. Abstract in *Science and Religion: One World – Changing Perspectives on Reality*, eds. Jan Fennema and Iain Paul. Enschede: University of Twente & Dordrecht: Kluwer Academic Publishers.

Hesse, M. 1981. Retrospect. In *The Sciences and Theology in the Twentieth Century*, ed. A. R. Peacocke. Stocksfield: Oriel Press.

't Hooft, G. 1992. *De bouwstenen van de schepping*. Amsterdam: Prometheus.

Isham, C. J. 1991. Quantum theories of the creation of the Universe. In *Interpreting the Universe as Creation*, ed. V. Brümmer. Kampen: Kok Pharos.

—1993. Quantum theories of the creation of the Universe. In *Quantum Cosmology and the Laws of Nature*, eds. R. J. Russell, N. Murphy, C. J. Isham. Vatican Observatory & University of Notre Dame Press.

Jastrow, R. 1980. *God and the Astronomers*. New York: Warner Books. (1st. edn. 1978, Reader's Library Inc.).

Kitcher, P. 1982. *Abusing Science: The Case Against Creationism*. Cambridge, MA: MIT Press.

Langford, J. J. 1966. *Galileo, Science and the Church*. Rev. edn. 1971 by Ann Arbor Paperbacks, University of Michigan Press.

Leplin, J. 1984. Introduction. In *Scientific Realism*, ed. J. Leplin. Berkeley and Los Angeles: University of California Press.

McFague, S. 1982. *Metaphorical Theology*. Philadelphia: Fortress.

McMullin, E. 1981. How should cosmology relate to theology? In *The Sciences and Theology in the Twentieth Century*, ed. A. R. Peacocke. Stocksfield: Oriel Press.

—1988. Natural science and belief in a creator: historical notes. In *Physics, Philosophy and Theology*, eds. R. J. Russell, W. R. Stoeger, G. V. Coyne. Vatican Observatory; distributed outside Italy by University of Notre Dame Press.

Meynell, H. 1982. *The Intelligible Universe: A Cosmological Argument*. London: Macmillan.

—1987. More Gaps for God? In *Origin and Evolution of the Universe: Evidence for Design*?, ed. J. M. Robson. Kingston & Montreal: McGill–Queen's University Press.

Monod, J. 1971. *Chance and Necessity*. Alfred Knopf. (Orig. *L'hasard et la nécessité*, Paris: Éditions du Seuil, 1970.)

Munitz, M. K. 1986. *Cosmic Understanding*. Princeton University Press.

Ovenden, M. W. 1987. Of stars, planets, and life. In *Origin and Evolution of the Universe: Evidence for Design*?, ed. J. M. Robson. Kingston & Montreal: McGill–Queen's University Press.

Pannenberg, W. 1985. *Anthropology in Theological Perspective*. Philadelphia: Westminster Press.

—1988. The doctrine of creation and modern science. *Zygon* **23**: 3–12.

Pedersen, O. 1988. Christian belief and the fascination of science. In *Physics, Philosophy and Theology: A Common Quest for Understanding*, eds, R. J. Russell, W. R. Stoeger, G. V. Coyne. Vatican Observatory; distributed by University of Notre Dame Press.

Penrose, R. 1989. *The Emperor's New Mind*. Oxford University Press.

Philipse, H. 1986. *Wijsbegeerte tussen twee culturen*. Leiden: Brill.

Polkinghorne, J. 1986. *One World*. London: SPCK.
—1989. *Science and Providence: God's Interaction with the World*. London: SPCK.
—1991. *Reason and Reality: The Relationship between Science and Theology*. London: SPCK
Prigogine, I. & Stengers, I. 1984. *Order out of Chaos*. New York: Bantam. (Transl. and rev. of *La Nouvelle Alliance*, Paris: Gallimard, 1979.)
Reeves, H. 1987. *L'Heure de s'enivrer, L'univers a-t-il un sens*? Paris: Éditions du Seuil.
Ritschl, D. 1984. *Zur Logik der Theologie: Kurze Darstellung der Zusammenhänge theologischer Grundgedanken*. München: Kaiser.
Russell, R. J. 1988. Contingency in physics and cosmology: a critique of the theology of W. Pannenberg. *Zygon* **23**: 23–43.
Sagan, C. 1988. Introduction. In S. W. Hawking, *A Brief History of Time*. New York: Bantam Books.
de Santillana, G. 1955. *The Crime of Galileo*. University of Chicago Press.
Surin, K. 1986. *Theology and the Problem of Evil*. Oxford: Blackwell.
Sutherland, S. R. 1984. *God, Jesus, and Belief: The Legacy of Theism*. Oxford: Blackwell.
Theissen, G. 1984. *Biblischer Glaube aus evolutionärer Sicht*. München: Kaiser. (Transl. as *Biblical Faith: An Evolutionary Approach*. London: SCM, Philadelphia: Fortress, 1985.)
Torrance, T. F. 1981. *Divine and Contingent Order*. Oxford University Press.
Toulmin, S. E. 1990. *Cosmopolis: The Hidden Agenda of Modernity*. New York: Free Press.
Whitehead, A. N. 1926. *Science and the Modern World*. Cambridge University Press.
Wildiers, M. 1982. *The Theologian and His Universe*, New York: Seabury Press.

# 10 The mind of God

Paul Davies

Of all the systems of thought aimed at understanding the world, what we call the scientific method stands out as the most successful. Not only has science led us to many new and unexpected discoveries about the world, it provides a powerful conceptual framework within which to organize our thinking about natural processes. Moreover, within the scientific community there is a remarkable degree of agreement about the way the world is.

In the chapters by Professors 't Hooft and Barrow we read that physics – the foundational science – may be approaching a point of culmination with a so-called Theory of Everything. We have been rightly cautioned against expecting too much from this endeavour; nevertheless, the very fact that such sweeping claims are even discussed itself attests to the power and scope of modern physical theory.

While remaining enthusiastic but sceptical about a Theory of Everything, I believe that the discussion of such a theory throws into sharp focus many of the basic assumptions that underlie the physicist's approach to the nature of reality.

In its most ambitious form, a Theory of Everything seeks to combine all physical laws and principles into a single, unified mathematical scheme, hopefully captured by a single, simple formula that you might be able to wear on your T-shirt. From this elegant and succinct source would flow a correct description of all the forces of Nature, all the basic particles and fields from which the material of the physical universe is composed, an account of the various masses, coupling constants and other parameters that describe how these particles interact, and a theory of space and time, including their dimensionality. But even this awesome list would not be sufficient to justify the name Theory of

Everything, for two reasons. First, such a theory would make no claim to explain many of the deepest mysteries that still confront science, such as the origin of life or the nature of human consciousness. Only if one adopts the stance of complete reductionism can an explanation of these sorts of phenomena be said to follow, implicitly and in principle only, from the physicist's Theory of Everything. At best, the physicist's Theory of Everything would account only for the elementary building blocks from which the richness and diversity of Nature derives. Second, a proper Theory of Everything, even in the aforementioned circumscribed form, also needs to include an explanation for the origin of the Universe as a whole before it can claim success.

Although I see no prospect for any of the Theories of Everything currently being discussed helping towards an understanding of the problem of complex systems, such as living and conscious organisms, I do believe that recent theoretical ideas point to a credible theory of the origin of the Universe.

There has existed since antiquity a deep problem concerning the origin of the Universe as a whole. If, on the one hand, the Universe has always existed, then the present state of the world cannot be fully explained by appealing to earlier states, because the chain of causation stretches back in time without end. We are faced with an infinite regress. On the other hand, if the universe came into existence suddenly at some particular moment in the past, then the first moment of time has a singular quality. What can one say about 'the join' between non-existence and existence? It seems inevitable that something beyond physical law – something supernatural – has to be invoked at time $t = 0$.

In recent years, a way has been found to pass between the horns of this dilemma using a variant of the big bang theory. The essential idea is that the Universe has not always existed, nor did it appear abruptly at one instant of time. Instead, it appeared gradually. By this I mean that time itself came into existence in a continuous manner. (Warning: 'gradually' here means extending over a Planck time. This is only $10^{-43}$ seconds. The big bang was still pretty abrupt by human standards!)

This new possibility arises as a result of quantum mechanics, and in particular the Heisenberg uncertainty principle, which admits an element of indeterminism in Nature. This implies that events, at least on a microscopic scale, can be truly spontaneous, i.e. occur without well-defined prior causes. This inherent uncertainty, or indeterminism, seems to provide a loophole for the Universe to appear without being 'created' or 'caused' in some specific physical way. Expressed differently, the spontaneous appearance of the Universe can be consistent with the laws of quantum physics. (In classical physics such an event would be miraculous.)

To summarise:

(1) Time (and space) originated with the big bang. There was no before.

(2) Given the laws of physics the Universe can bring itself into existence. More precisely, the appearance of a Universe from nothing need not violate the laws of physics, once the existence of quantum phenomena is taken into account.

However, the Universe is not only self-creating, it is self-organizing too. From what is known about the very early stages following the big bang, the Universe was essentially featureless. Indeed, it may have consisted of little else than expanding empty space! All the rich diversity and complexity of physical forms and systems that adorn the cosmos today have arisen since the beginning in a long and complicated sequence of self-organizing and self-complexifying physical processes. No imprint of this richness was stamped on the Universe at its birth. It has emerged as a consequence of the inherent creativity of Nature.

These features undeniably give the Universe the appearance of design and purpose. It *looks* as if it is directed towards some goal. Let me first say something about design. All design arguments in Nature proceed by analogy with human design, and are open to interpretation and opinion. We have different thresholds of amazement in the face of the subtlety and ingenuity of Nature. Nevertheless each of us would, presumably, at some level of evidence consider Nature to have been intelligently designed. As the philosopher John Leslie has remarked, if we habitually came across rocks with 'Made by God' stamped on them we would be unlikely to dismiss the world as a random accident. A better example might be that if we were to sequence the human genome and discover it contained an intelligent message for us we could reasonably conclude that our existence was not totally fortuitous. These are probabilistic arguments to do with evidence, and we make such judgements all the time.

Regarding purpose, it is often said that this is a purely human concept; that it is meaningless for the Universe to be said to have a purpose. Again, it is a question of when we can accept the analogy with human affairs. Most people are perfectly content to proclaim: 'The Universe is a machine!' because mechanistic thought has dominated science for two hundred years. Yet the Universe clearly isn't a machine. True, it has mechanistic aspects to it, and the machine concept captures *some* element of the nature of the cosmos, but it is hardly the whole story. Recently a number of scientists such as Ilya Prigogine have emphasized those aspects of the Universe that are reminiscent of living organisms, so that one may now reasonably use words like 'organization', 'adaptation', and 'behaviour' in connection with some non-linear complex inanimate systems. Now living things have purposes, so to the extent that the Universe resembles a living organism (or, more plausibly, an ecosystem) it may be said to be purposeful. Of course the Universe is *not* a living organism any more than it is a machine. Yet this 'lifelike' aspect does capture some part of its nature, and a concept analogous to purpose may therefore apply just as well as the concept of mechanism.

I should like to tease out a subtlety concerning purposefulness in the Universe. Some years ago I wrote a book called *The Cosmic Blueprint* where I wanted to draw attention to the fact that the laws of physics are like a blueprint for the Universe – an algorithm might be a better word. There is in some sense a cosmic blueprint because these laws of physics bestow upon matter and energy an uncanny ability to organize themselves, a disposition to arrange themselves and evolve along certain pathways leading from simple to complex. This blueprint is not, as some people would mistakenly believe, so rigid that it fixes everything in detail. It's not as if there exists in some Platonic realm out there a complete plan for the Universe – including the colour of my eyes and the words I have written. I don't think that is how the laws of physics operate at all. The future is open in its details. We must reject rigid determinism. But there is a tendency or disposition for matter and energy to develop towards more and more complex states.

I like to call this way of looking at the world 'teleology without teleology' because it's really telling us that the laws of physics are special in form and they are such as to facilitate the emergence of certain complex structures from a featureless origin. This subtle interplay between chance and necessity of the laws – which are very special in form such that they permit the emergence of complexity and yet allow a certain openness in the way things develop – is very difficult to capture in a succinct phrase. Charles Birch has expressed it well: 'creation is not by fiat but by persuasion'.

A useful analogy is with the game of chess. The rules of chess are fixed in advance. They are chosen to facilitate a rich and interesting variety of play. But the rules alone do not determine the specific outcome of a particular game, merely the general trend of play. There is an analogy here with the laws of physics. The laws are 'there', and they encourage the emergence of richness and organized complexity from initial featureless simplicity, but not with a rigidity that fixes all the details in advance. The 'fixing' of the laws of physics to have this special 'richness-facilitating' property is sort of teleological in its overall effect, but there is no specific teleological 'pull' in a given evolutionary cosmic history. Actually, a better analogy is with the cellular automaton game by John Conway called *Life*. Here the rules of the game are again chosen so as to encourage the spontaneous emergence of self-organizing complexity from random input.

To summarize, then, the remarkable nature of the laws of physics. These laws:

(1) Permit the Universe to come into being from nothing;
(2) Encourage it to self-organize;
(3) Fix its evolution in outline (e.g. from simple to complex) but not in detail;
(4) Bestow upon the Universe the appearance of design.

This does not exhaust the list of remarkable properties. In addition, these laws are *intelligible* to us. Through science, we can come to understand the Universe. What impresses me is that not only can we make sense of this or that little feature of Nature, but given the recent major advances in physics and cosmology we actually seem to be able to provide a convincing explanation for a very wide range of natural phenomena. Indeed, some of the more optimistic among us talk about a Theory of Everything.

Throughout history there has been a dream that mankind would one day come to know the reason for the existence of the Universe. Some believed that this ultimate knowledge could be attained through mystical revelation. Others supposed that rational reasoning would provide the key. In the modern era, science has been seen as the natural route to this compelling goal. Most of the time, however, the hope that we might be able to unlock the ultimate secret of existence has been dismissed as a vain search for a chimera. But in the last few years an increasing number of scientists have given their attention to this topic. They have asked whether an ultimate explanation for the physical world is possible, and if so what form this explanation might take.

All discussions of ultimate explanations are predicated on the assumption that the Universe is, at rock bottom, rational, and therefore can be understood by systematic investigation and human reasoning. Not all cultures, however, share this view. Many ancient communities believed that Nature was under the control of capricious gods who exercised arbitrary and unpredictable power. Other cultures have regarded the Universe as fundamentally mysterious, with no ultimate principles of order.

Science emerged from mediaeval Europe, under the twin influences of Greek philosophy and the Judaeo-Christian religion. The Greek philosophers were deeply influenced by the power of systematic human reasoning to make sense of the world. They believed that men and women could come to know the nature of the Universe through logical thought. Some, such as Pythagoras, were convinced that the Universe had an underlying mathematical nature, and that the development of mathematical theory would provide the key to unlock all cosmic secrets. We can see signs of how closely the classical world identified number and geometrical form with the fundamental logicality of the cosmos in our words 'rational' and 'ratio'.

Meanwhile, the Jewish religion established the concept of a transcendent God who created the Universe at a certain moment in time, and ordered it by imposing laws. There were laws for men and women, and laws of Nature. The world was seen to possess a distinct historical progression, starting with the creation, and unfolding towards a final state. The processes of Nature were regarded as part of God's directed plan. This powerful image of God the omnipotent lawmaker was subsequently incorporated into Christianity.

For some centuries these two strands of thinking remained separate. Then, in the thirteenth century, the works of Plato and Aristotle were rediscovered in Western Europe, and their fusion with Christianity created a potent conceptual mix. A young Dominican friar, Thomas Aquinas, began applying the rigorous techniques of Greek geometry, with its axioms and theorems, to the study of theology. Aquinas conceived of a perfect and rational God who created the Universe as a manifestation of his supreme powers of logical reasoning. A key feature of Aquinas' system is that God the Lawgiver exists outside of time: his laws are therefore transcendent eternal truths after the fashion of Greek mathematics.

Although Aquinas' image of God is highly abstract and remote, it dominated Christian, hence European, thinking for several centuries. When Newton and his contemporaries formulated the basic principles of physics in the seventeenth century, they were convinced that their discoveries revealed God's handiwork, and that Nature's rational order had a divine origin. These scientists conceived of a universe ordered according to definite laws of Nature, which they treated as 'thoughts in the mind of God'.

Subsequent generations of scientists accepted the idea of the laws of Nature as timeless eternal truths, but their theistic provenance was gradually cast aside. The laws became 'free-floating', self-supporting principles of regulation and order that were simply accepted as 'given'. The laws themselves acquired some of the qualities that were formerly attributed to the God from whence they were once derived. Thus we find today's scientists generally agreed that the fundamental laws of physics are universal, absolute, omnipotent and eternal. Many believe them to be transcendent too, in the sense that the laws exist independently of the state of the physical universe, and possibly independently of the Universe altogether.

Very few scientists, however, stop to wonder where these laws come from, or why they have the form they do. In particular, they do not question the fact that we human beings are apparently capable of knowing these laws. Yet the entire scientific enterprise is founded on the assumption that the Universe is rationally ordered and that we, as rational beings, can come to know that order through the processes of reasoned enquiry that we call science. For Newton and his followers this was no mystery. They believed that Man was created in God's image, and so shared his divine rationality, albeit in diminished form. Thus Man is able to discern this same rational structure reflected in Nature, because the physical world is also a product of God's rational plan. But if the laws are no longer regarded as thoughts in the mind of God, the success of science seems baffling.

The mystery is all the greater in that the laws of nature are not at all obvious to us on casual inspection of Nature. The simplc law of falling bodies, for example, was discovered by Galileo only after careful experiment and observation, and was greeted with general scepticism. Most people cannot 'see' intuitively that heavy and light bodics accclerate equally under the Earth's gravitational force. Indeed, this law is often masked by the complicating effects of air resistance. The underlying order in Nature is therefore hidden from us, and must be deduced by often elaborate procedures. The late American physicist Heinz Pagels expressed this hidden quality by referring to 'the cosmic code'. The laws of nature, he would say, are written in a sort of secret code, and therefore we do not perceive them directly. The job of the scientist is to 'crack' the cosmic code and read the 'message'. Scientists do this by a careful combination of experiment and theory. Experiment is regarded as an interrogation of Nature, from which the scientist receives cryptic answers, and the theorist then tries to 'decode' these answers and fit them into a rational scheme.

The problem, for the atheistic scientist, is to understand how this code-breaking ability of human beings has arisen. If our mental as well as our physical attributes are the result of biological evolution, then the ability to deduce the laws of nature ought to have some selective advantage. However, it is hard to see what this advantage might be. It is sometimes argued that the ability to dodge falling objects, jump streams and spot natural rhythms confers certain advantages. But these feats are accomplished instinctively, not cognitively. They are shared by many animals that have no understanding of natural laws. The ability to perform such tasks is programmed into the brain on the basis of previous experiences with similar situations. We do not use our reasoned understanding of the laws of physics, for example, to avoid a falling tree. Furthermore, the most spectacular successes of science apply to atomic physics and astrophysics. Unravelling the secrets of the atom or figuring out the structure of a black hole are scarcely relevant to survival in the jungle. Our brains, it seems, are remarkably fine-tuned to spot the patterns and order of Nature in domains that are quite irrelevant to biological evolution.

A study of how science works only serves to deepen the mystery. Most of what is known as hard science, such as fundamental physics, is cast in mathematical language. All the basic laws of physics can be expressed in succinct mathematical form. Thus the dream of the ancient Greek philosophers – that the world is a manifestation of mathematical principles – remains alive in modern physical theory. Of course, today's mathematical physicist goes well beyond Euclid's geometry in providing a mathematical scheme to describe Nature. Modern branches of mathematics such as group theory, differential geometry and topology are needed too. So deeply does mathematics penetrate the principles of the Universe that the English astronomer Sir James Jeans once proclaimed: 'God is a pure mathematician!'

Yet mathematics is not given to us: it is a product of the higher human intellect. It stems from the more developed parts of the brain – the most complex system known. How odd, therefore, that this most highly processed system should create something that applies stunningly well to the most elementary processes of Nature, such as subatomic physics. Again, there seems to be little biological advantage in humans having the ability to perform calculus or solve differential equations. Yet this ability does not have the appearance of a meaningless quirk of evolution. Mathematical skill is widespread in the population. Moreover, there sometimes occur mathematical geniuses of breathtaking accomplishment, such as the self-taught Indian clerk Srinivasa Ramanujan, who wrote down large numbers of correct mathematical formulae seemingly without effort, formulae which sometimes needed years of work by some of the world's finest mathematicians to prove. The fact that such geniuses occur in every generation shows that exceptional mathematical ability is a rather stable feature of the human gene pool. But why?

In recent years mathematical physics has been dominated by the goal of unification. The hope and expectation of many theoretical physicists is that all the fundamental laws of physics can be merged into a single superlaw or unified theory. This theory would be encapsulated in a succinct mathematical expression which, it is sometimes quipped, you could wear on your T-shirt. From this all-embracing formula a description of all of Nature would flow. It was the prospect that such a theory might be attainable in the near future that led the mathematician Stephen Hawking to entitle his inaugural lecture 'Is the end in sight for theoretical physics?', on receiving the Lucasian Chair (the chair once held by Newton) at the University of Cambridge.

Now although this hope for unification may be highly optimistic, it is certainly the case that in the three short centuries since Newton, science has advanced to the point where we have available mathematical theories that accurately describe an enormous range of physical phenomena. Indeed physicists believe that we can give a good explanation for most physical processes at the fundamental level. The theories of the four basic forces of Nature plus quantum mechanics hold up admirably across a very wide range of experiment and observation. Furthermore, these theories enable us to understand not only the workings of the microworld within the atom, but also phenomena on a cosmological scale. Current physical theory thus encompasses, albeit in a provisional way, an accurate description of the world from the smallest to the largest scales of size.

Perhaps scientists have become so blasé about this success that they do not stop to wonder at how extraordinary it is. The laws of the Universe could so easily have been impenetrably complicated or subtle, and we might never have come to know them. Yet it appears as if these laws, while cryptic and esoteric, are nevertheless captured by a level of mathematics that is within human capabilities. Let us take a closer look at this happy circumstance. In practice, mathematical physicists are usually able to engage in basic research only after about twenty years of education and training. History shows that most major advances in this subject are made by young scientists, perhaps in their twenties or early thirties. It therefore seems as if the necessary education span is rather close to the productive life span, itself a significant fraction of the biological life span of humans. In other words, human beings can only just achieve the necessary level of mathematical creativity to crack the cosmic code. But factors such as the education span and life span of humans are purely biological qualities, presumably related to selective effects in evolution. What possible connection can they have with the mathematical complexity of the laws of physics? How strange then, that humans find themselves just able to grasp this complexity, and not merely for some limited range of phenomena, but on a truly cosmic scale.

Some people may shrug these facts aside as a lucky but meaningless coincidence. In my opinion, however, they point to a deep connection between our existence as conscious, thinking beings, and the existence of the physical universe, with its various laws and systems. I don't mean to imply that *Homo sapiens* as a particular species is special, only that the emergence of *mind* as a phenomenon in the Universe, somewhere and at some stage, is a fundamental and not an incidental feature. I draw this conclusion because it is clear that the higher-level phenomenon of mind links up in a profound way with the bottom-level structure of the physical world, i.e. the basic laws and particles of nature.

Recently the Oxford physicist David Deutsch drew attention to a curious aspect of this linkage. We normally think of mathematics as a vast body of formal relationships. Although in practice human beings are able to carry out only a tiny proportion of all computations, there seems to be a widespread belief that, given a large enough computer and a long enough run time, all possible mathematical operations can in principle be carried out. In fact, this is quite wrong. In the 1930s the logician Kurt Gödel demonstrated that there exist *undecidable* mathematical propositions. Roughly speaking, this means that there are mathematical truths that cannot be proved to be true. Furthermore, these undecidable propositions are not restricted to rare and abstract branches of mathematics; they exist even in basic arithmetic. Shortly after Gödel published his proof, the English mathematician Alan Turing (co-inventor of the computer) adapted it to show that there exist uncomputable numbers. These are numbers that can be proved to exist, but which cannot be computed by following any systematic procedure or algorithm, of the sort used by electronic computers for example. In a certain sense, only a tiny portion of the elaborate body of mathematical truth is actually computable.

Now Deutsch points out that the operation of a computer, and by extension the human brain, depends on the structure of the physical world, and in particular on the actual form of the laws of physics. One can imagine different universes with different laws and structures in which mathematics that is computable in our world would not be computable there, and vice versa. In other words, what is computable is decided by the laws of physics. But we have also seen how human – hence computable – mathematics provides a mysteriously succinct, indeed simple, description of the laws of physics. So we glimpse here a closed loop of consistency. The laws of physics give rise to a world in which certain mathematical operations are computable, and these same operations encode in simple and elegant form those very laws of physics that permit the computability.

The question obviously arises of whether this self-consistent loop is unique. Might our world be the only possible one with a computable cosmic code? If so, then our existence as part of this computation-cycle is surely of the deepest significance. If there are other self-consistent worlds, would they permit the emergence of complex structures such as biological organisms that would become aware of those worlds, or is ours unique in that respect? Although we do not know the answers to these deep metaphysical questions, I personally believe that the close matching of mathematically skilled thinking beings to the mathematical structure of the physical world they inhabit is so improbable that it must be unique.

The mathematical self-consistency of the physical world is only one strand of evidence that mind is a fundamental rather than incidental feature of the universe. There are also the curious 'coincidences' that go under the name of the Anthropic Principle. For some time scientists have been struck by the fact that our existence as complex biological organisms depends very delicately on a number of apparently fortuitous accidents of Nature. It seems likely, for example, that if the laws of physics were even slightly different in form then many other structures, such as stable hydrogen-burning stars like our sun, would not exist. A systematic study reveals that in a universe with slightly modified laws the conditions necessary for biological organisms to arise would not occur. Only in a universe with laws and conditions rather similar to our own would conscious beings emerge and ask about the meaning of existence.

It has been conjectured that the laws of physics in our universe may be in some sense 'optimally coded' to produce a rich variety of complex forms and systems. Gottfried Leibniz suggested that we live in the best of all possible worlds. Some scientists have suggested that this idea might be made mathematically precise, to reveal a sort of Principle of Maximal Variety. According to this principle, the laws of physics have a particular mathematical form that encourages the emergence of complex organized systems such as biological organisms, and, ultimately, thinking beings.

It may be that rapid scientific progress is just a quirk of twentieth-century history, and that we are as far as ever from knowing 'the mind of God'. Yet I feel compelled to believe that, through science, we really are being allowed to glimpse the rational foundations of physical existence. This privilege – that we are privy to large parts of the cosmic code, and may even come to know it all – seems altogether too good to be an accident. Freeman Dyson once wrote in this context: 'I do not feel like an alien in this universe. The more I examine the universe and study the details of its architecture, the more evidence I find that the universe in some sense must have known that we were coming'. I feel the same way too. In some strange and perhaps unfathomable manner, it begins to look as if we are meant to be here.

# 11 The sources of models for God: metaphysics or metaphor?

Mary B. Hesse

## Introduction

In the last three centuries physicists have come up with a surrogate for God. It is the conception of laws of Nature as universal and eternal, comprehensive without exception (omnipotent), independent of knowledge (absolute), and encompassing all possible knowledge (omniscient). In other words the structure of physical laws has all the classic metaphysical attributes of the Deity (Davies, 1992). In this light are discussed many of the perennial problems about the meaning of the Universe that have traditionally been the concern of religion: how and when the Universe began, possibly how it will end, its evidences of design, its deep and comprehensive orderliness, the beauty of the theories devised to explain it, the wonder of human capacities to penetrate its mysteries, the duty that rational beings have to pursue precise explanations of things as far as these will go.

There is no doubt that there is great fascination and excitement among the educated public about these metaphysical- and theological-sounding claims. Why? Probably for two chief reasons: first because in our secular society we have lost the depth, coherence and value that used to be given to life and the world by religion, and second because natural science has appeared to be the only decisive successful way of reaching the truth about things, and has been exploited as such by post-Enlightenment philosophy, particularly that of the Anglo-American world. 'Lord, to whom shall we go, thou hast the words of eternal life,' was said to Jesus in the New Testament; it might now be said to the spirit of natural science as our society's lord.

And yet ... everywhere in so-called post-modern thought this philosophy is breaking down. The very pragmatic 'success' of science is questioned when so many of its outcomes seem to be out of control: nuclear energy, the greenhouse effect, even some spin-offs of medical science. It is noticed that many alternative ideologies and belief systems have functioned more or less successfully in other societies, and these societies are now studied in depth by social historians and anthropologists. Even when distant in time and space these alternatives often seem to have lessons for our own technology-dominated lives. Knowledge of them leads to intellectual confusion of values and the collapse of the delicate social balance between the modern and the traditional, with consequences becoming all too apparent everywhere in the developed world. In a sort of disseminated but bastard Marxism that would not have been approved by Marx, many have concluded that truth and values are internally defined by each society, and that there are no universally valid principles for the ordering of human life. In currently popular sociology of scientific knowledge even scientific claims to understand the natural world are subjected to the same relativism: it is argued, sometimes quite cogently, that scientific theories owe much to surrounding ideology and social and economic conditions. All seems to be at the mercy of social power groups and their interests, or of professional whim and fashion.

Where such relativism has been contemplated and recoiled from, it is not surprising that there is a reversion to scientific realism in the attempt to build on what is still perceived as being the only reliable route to knowledge. The modern realist ideology of natural science even culminates in the claim that there is in principle a Theory of Everything, which would provide a conclusive answer to all the perennial religious questions, by showing that there is no logical space left for realities beyond the natural world. Other contributors to this volume discuss this claim in detail, and conclude that what are currently called 'Theories of Everything' in cosmology refer only to fundamental physical laws, and should not properly claim to exhaust all particular or contingent features of the world such as are of immediate relevance for human life. The hairs on our heads are not all numbered in a Theory of Everything, even if this could be made complete and unique as a physical theory.

Many cosmologists do, however, look to the fundamental physical laws as being related in some way to the concept of God. But I believe there are cogent reasons why the attempt to build models for God from the structure of natural laws is misguided. I shall first give some reasons for thinking that the strong concept of 'laws of Nature' that is implicit in these cosmological speculations is mistaken, and replace it by a brief statement of a 'moderate realism' which seems to me to be all that the facts about natural science justify.

### Moderate realism

The concept of laws of Nature originated in a seventeenth century legal metaphor, according to which God's rule in the cosmos was likened to human authority in the state. There was however the important difference that since God's ultimate intentions are inscrutable, we have to come at his laws of Nature indirectly, by careful and objective scrutiny of the accessible facts, that is to say by the new experimental method of 'putting Nature to the question' and setting up experimental situations which will force Nature to give relatively unambiguous answers. That the method could be successful (quite quickly, they believed) was further guaranteed by a belief in the 'two books', that is God's word in Scripture and in Nature, and that Nature is written in an alphabet which must be decoded. Letters and words of the alphabet correspond to real kinds of objects in the world: 'atoms' to atoms, 'light' to light, and the sentence grammar of the natural language corresponds to the laws of Nature relating objects to each other. This was the origin of the subsequently entrenched realism and the correspondence theory of truth, in which natural science was supposed to discover real hidden structures of the world, and to demonstrate them as constituting a necessary causal network that even God cannot override. But of course soon after the seventeenth century this scientific metaphysics began to lose touch with its essentially theological grounding.

I suggest that recent philosophy of science has undermined most of the tenets of this metaphysical theory. The belief in necessary laws of Nature is closely related to the belief that there are natural *kinds* or species of objects which allow us to interpret particular events, which are all we can observe, as mere instances of underlying universal laws relating objects to each other just in virtue of the kinds of objects they are. We 'intuit', as Aristotle said, the universal in the particular: one egg can stand for all eggs. This is just the metaphysics rejected by Hume in favour of his view that objects are all loose and separate, and that therefore we have no rational justification for passing from the particular to the general, only habits according to which we have learned as far as possible to anticipate what the natural environment throws at us. 'None so like as eggs' he says, yet they are in detail different, and this is sufficient to make the inference to a natural kind and its accompanying laws logically invalid.

The question whether Hume or Aristotle was right about laws, then, is bound up with the question of the existence of natural kinds, particularly natural kinds of objects which are not directly observable, but are postulated in theories to explain observables. These fundamental kinds (for example in physics they are hydrogen, oxygen, atoms, electrons, photons etc.) are held to be explanatory, in terms of universally valid laws, in so far as they capture the relatively simple structure of Nature underlying the medley of complex and mutually interfering phenomena we directly observe. But are there natural kinds? In recent historical studies we have become used to the idea that the theories which picture these fundamental entities are subject to revolutionary change through the history of science, and that there is no sign of that convergence upon the basic alphabet of concepts that was held in the seventeenth century to be the goal of experimental science.

Let us take some examples. Consider 'electrons'. Are they 'real'? We know a great deal about small objects that have a certain measurable charge and mass, instances of which are identifiable and distinguishable from other types of micro-particles in a great variety of experiments. 'Electrons exist'. Everything we know experimentally about electrons has become part of the deposit of positive science, but the fundamental theories of micro-physics are controversial and at present highly volatile. Nothing can rule out the possibility of a 'revolution' in physical theory that will re-interpret many identifications and laws about electrons, in a way similar to what happened, for example, to phlogiston. There are two aspects to such revolutions. First they *leave alone* many local and particular truths, where we may as well say we are referring to 'electrons', even though in a new theory electrons may not have all the apparently essential characteristics we now postulate of them. Hydrogen goes up in flames, even if Priestley called it 'phlogiston' in referring to experiments that he did and we are able to repeat. Electrons by any other name activate cathode ray tubes and Geiger counters. But in a different sense the new theory may show that 'electrons do not exist', because our present theory about what they are and how they behave is found to be in detail and in general false. In just the same way we now say 'phlogiston does not exist', that is, phlogiston is not a natural kind, because Priestley's theory is in detail and in general false. It is false of the truths that have now entered experimental or common-sense realism.

The local and particular success of science does not require there to be true *general* laws and theories, even about common-sense objects, or at least it does not require us to *know* them. We are always in local and particular situations, however far into the past or future or into the micro- or macro-scale our theories claim to penetrate. Moreover, our application of general-kind terms to common-sense objects, as well as to scientific objects, is always subject to ambiguity of meaning when extrapolated to distant conditions. We are trapped in some natural language or other when we talk of the real world, and because we never know whether the general-kind terms in that language really represent natural kinds, we do not even know how to express the properties of objects under highly extrapolated conditions. Local meanings and local laws are contextual, and work without having to capture natural kinds or essences. And without natural kinds we cannot express true general laws.

So far we have noted two features of what may be called a 'moderate realism' for natural science. First there is the *plurality of possible theories* of fundamental kinds, and the impossibility of reaching, or at least of knowing, any ideal theory with 1:1 correspondence to the world. Second, science justifies only *local lawlike relations* between objects, not strictly universalizable causal laws. But of course a third feature of natural science has been presupposed throughout, which explains the apparent fact that science is in some sense *objective* and *progressive*.

This feature is the requirement that predictions from postulated laws and theories should be satisfied by the results of observation and experiment. Science has undoubtedly been successful in ordering the (always local) natural environment to permit prediction and control. This is what constitutes 'progress' in science, not the discredited idea of convergence of theories to 'truth', and it seems to me to be the overriding cognitive value of natural science as this has been institutionalized since the seventeenth century. It is an instrumental criterion, but it does not prevent us interpreting theories as descriptive of local lawlike structures of the world (regularities, measures, symmetries), though not as universal natural necessities, nor as the ultimate 'what-is' of the natural world. That is what makes this realism 'moderate' (Hesse 1992).

## Cosmological models for God

So much for general objections to the realist theory of laws, which are drawn partly from historical studies. But it may be replied by cosmological realists that physics has now passed a kind of historical threshold in which we have finally got it right, or approximately right, and that there will not in future be the sort of 'Kuhnian' revolutionary change in theory that occurred in the past. In particular, it will be said, modern mathematical physics is interested not so much in the substantial entities that form the building blocks of the Universe, but in mathematical structure, and this remains invariant to changes of the picturable models that may be used to aid its understanding. The metaphysical God of Paul Davies seems to require at least this much mathematical realism. There is some force in this reply, and a moderate realist of my kind therefore has to argue that such mathematical structure, even if convergent on true laws, is not strong enough to serve as a surrogate model for God.

Firstly, the models are purely formal and static, and have no qualities, especially personal qualities, which touch the religious problems of everyday life. Secondly, they do not introduce any values other than those of orderliness, mathematical depth and elegance, particularly not goodness nor human freedom. It follows that they do not address even those paradoxes that have been found in traditional metaphysical theories of God. For example, if he is omnipotent and omniscient, and also 'good', how can he permit evil? Or if he is eternal, and past, present and future are before his eyes, how can we be other than puppets bound to perform according to his all-encompassing knowledge? How, in other words, can we make sense of human freedom?

My strategy now will not be to go into details about the pros and cons of different models of God, but rather to lay out a few alternatives to the cosmological model, and give some indication of the sort of arguments that may be used for and against them. Most if not all of these arguments will turn out to be quite different in character from those deployed in the natural sciences, particularly because they will introduce value-judgements. I shall conclude in the last section that they should all be understood as better or worse *metaphors* for God, not metaphysical descriptions of him. First I suggest some alternative sources of models.

### Evolutionary models for God

One objection to cosmological models is that they are essentially static. But even physics introduces the reality of *time* into some of its theories, and evolutionary biology reinforces this dynamic picture. Because biology recognizes contingency and probability among its causes, it is not comprehensively lawlike, and therefore allows for contextual explanations and open unpredictable futures. For those who equate explanation and understanding with the *absence* of contingency, this is a temptation to 'fill the gaps' with some metaphysical or teleological explanation, such as that God is an agent who directs the Universe along just one of the possible paths allowed by evolutionary science. God then becomes the designer of some of the desirable possibilities that have been realized such as the apparently contingent evolution of life-forms in a supportive environment, and the appearance of human beings with their rational, aesthetic and spiritual potentialities. On this view, of course, God has to be reckoned the designer of undesirable features as well, such as the capacity of the world for evil and disaster whether natural or created by humans.

An alternative and better evolutionary view is that God is the goal of the Universe – the teleological state to which evolution tends, irrespective of evil contingencies and meaningless and valueless dead-ends. Such a model usually presupposes that the goal of the Universe will have some relation to what we would antecedently regard as 'good'. But we cannot predict the future of actual evolution to ensure that trends we regard as desirable will continue. On the other hand, we cannot just *define* the future as the good, without taking the risk of violating our own moral insights. A model of a good God has to presuppose a moral structure of the Universe which can certainly not be derived from the scientific facts alone.

These evolutionary models of God exploit the emphasis of evolutionary science on dynamism, progressive change, and in general the meaningfulness of time. But these features are by no means new. They are often seen as the great insight of post-Newtonian science: the legacy of eighteenth- and nineteenth-century cosmology, geology and biology, and of the romantic movement. But they should rather be seen as a common-sense return to the historical consciousness of the Judaeo-Christian tradition, rejecting not only the aberrations of Newtonianism, but also those of Greek metaphysical rationalism. The biblical God on the contrary was always the historically active God of both Nature and the human race, a God who is the origin of value as well as fact, and hence who is concerned with the right ordering of human life beyond its merely material basis.

### 'The starry skies above and the moral law within'

Kant drew as a necessary conclusion from the reductive science of Descartes and Newton that considerations of morality and goal-directedness must be distinguished from those of natural philosophy. The practical and the teleological reason have reasons that pure reason (what we would call 'scientific' reason) knows not of. It is significant that in the English-speaking world, on the whole it is Kant's *pure* reason that has been exhaustively studied, to the neglect of his philosophy of ethics and religion, and hence we deprive ourselves of his argument that reason goes beyond natural science.

Kant has, however, bequeathed us two unfortunate presuppositions for the study of morals and religion. One is his rationalism, the other his individualism. His theories of ethics and religion rest on the same rational pattern as his theory of scientific knowledge, namely the search for that which we cannot but presuppose in order to function in the way we do function. The laws of substance, causality and interaction are assumed to be given in our ordinary commerce with Nature, similarly the moral law and its duty, and the assumption of a 'kingdom of ends' are given as universal a prioris in our practical and goal-directed activities. This view is of course highly problematic. Related to it is Kant's subjectivist psychology. It is the individual Self who has this knowledge, operating as an independent atom in the flux of events. But this form of individual rationalism has I think now become untenable in the light of further developments, primarily in the sociology of knowledge.

### 'Society is God'

One of these developments leads, through the work of Durkheim on the sociology of religion, to a *social* in place of an individualistic model of God. Here the phenomena of religion are studied among the facts which are intrinsically social, and not reducible to the psychology or intentionality of individuals. In the sociology of groups, religions are distinguished as being concerned with the *sacred* – things set apart, yielding symbolizations in terms of which profane objects acquire power to unify communities. Thus arise permissions and prohibitions in relations between individuals and Nature, and mutually between individuals, which act as external constraints upon belief and practice, just as do natural facts. Durkheim argues that there must be some source outside the individual for such constraints, and finds it in society itself, felt at first as a reality barely separable from the reality of the natural world. A brief and misleading summary of his thesis is 'Society equals God', but consistently with this he insists that religion is not reducible to other social forms, since it has its own autonomous existence as the space of sacred symbols. In particular, religion is connected with the ethical structure of a society, for which it provides symbolic social sanctions and psychological motivation.

This model of God has the merit of emancipating religion from excessive individualism and also from a pantheistic identity of God with the natural world. In these respects it is nearer than the scientific models to an adequate account of the phenomena of the world religions. But it also, according to Durkheim, emancipates religion from any transcendental reference. He finds no need for the hypothesis of a transcendent God outside society, for he thinks society alone is sufficient to explain the observable religious phenomena. This is not a view likely to be accepted by most religious believers, whose religious practices and experiences usually presuppose belief in some extra-natural referent for the concept of God.

There are other fundamental difficulties about Durkheim's theory. One is his general neglect of the relative autonomy of many of the religious experiences of individuals which do not appear to have 'society' as their objects. Another is his tendency to reify 'society' in ways that he finds philosophically objectionable in connection with the concept of 'God'. He also has an exaggerated faith in the possibility of positivist social science, and particularly in evolutionary justifications of moral belief. Like some later sociobiologists, he tends to use such justifications as standards of social progress and social good.

### God as 'personal': the God of the gaps?

The transcendent God that Durkheim rejects is the sort of personal God that is the predominant model of pre-modern mythology as well as of most monotheistic religions. But note that this as such is not the *Christian* model, which is that of 'Three persons in one God', meaning three *personae*, or masks, of God. Whatever the difficulties and controversies that have attended this conception throughout the Christian era, one thing at least is clear: this is a metaphoric way of speaking, and to understand it we have to concern ourselves with the nature of metaphoric language in general, and in particular how we can meaningfully use metaphors of God.

Before considering this in the final section, it is worth mentioning at least one of the difficulties in reconciling a 'personal' God with science, if he is to be understood as a supernatural actor in history. The principal difficulty concerns the antithesis of determinism and chance. Physics describes the world either in terms of a wholly deterministic structure of laws in which there is in the last analysis no room for contingency, or in terms of fundamental chance occurrences which are themselves subject to statistical laws, so that any outside 'intervention' appears to conflict on the macro-scale with the discovered distributions of probability. A determinist world seems to rule out all possibility of spontaneous action, whether by an external God, or by human exercise of free will. The consequent personal model of God in the seventeenth century was the Deist God, who started the Universe going with its fixed laws and initial conditions, and was powerless to influence events thereafter. The modern physical versions of a closed and determinist universe would appear to rule out external influence altogether.

The physical alternative is to accept fundamental chance in the context of statistical laws of physics. This model seems to leave the logical possibility of a 'god of the gaps', with divine and human free action in particular single events, while not denying the big overall probability distributions at the macro-level. But to regard God's actions as random particular events is a purely otiose hypothesis from the physical point of view, and does nothing in itself to give a positive account of them.

On the other hand, autonomous personal action, whether human or divine, and whether made in terms of a determinist or indeterminist universe, cannot logically be ruled out. Objections to this model must rest on the very strong presupposition that all natural events are reducible to the laws of physics. This presupposition need not take the form of a 'Theory of Everything', but may be the weaker and more venerable claim that the phenomena of all physical systems, of biology and of human physiology and psychology, are in principle explicable as special cases of physical law. This was the explicit assumption of Laplace when he derived a determinist world from Newton's mechanics. But in spite of 300 years of trying, we are surely no nearer justifying this claim than were the seventeenth-century mechanists, and many contemporary biological and human scientists would themselves want to reject it in favour of more open and flexible scientific models. If we have reached an impasse in trying to find models for the notion of the 'self' required to understand free will, or the notion of 'intervention' required for a personal model of God, we should surely look initially not to physics, but to these other sciences.

All this does no more than make a space within physical science for a model of the active and intervening God. How we should give a positive description is a different and more complicated matter. So finally I will look at the classic attempt by a master of his contemporary science and theology, Thomas Aquinas.

## What all men call God

The crucial question as Aquinas saw it was 'How can we speak of God the eternal creator *at all* in human language?', and, if we can, 'What are *good* models for God and why?'. Aquinas stood within the Judaeo-Christian tradition that God is primarily a moral rather than a metaphysical entity, to be known by the practical rather than the pure reason, and that he is to be described in personal as well as metaphysical terms.

Aquinas is best known for his so-called 'Five Ways' to 'prove' the existence of God, all of which in some sense depend on evidences from the observable world. He argues that since all things change and come into being as effects of other things that cause them, there must be an uncaused first cause. Again, all observable things are contingent (might not have existed), so the first cause must be a necessary being; all things are observably imperfect, so need a perfect cause, and all things must have a purpose or design or goal, so the first cause is also the final cause and goal of the creation.

Wrapped up in Aristotelian metaphysics of causality as all this is, we need not enter into the structure of the arguments. We can of course find in the modern cosmological theologies formal parallels for the quasi-divine attributes of a necessary, ordered, perfectly accurate, causal network of physical law, and even for the argument from design. Some of the mathematical properties discovered in physics may seem to mimic the attributes of God, but as I have argued they are purely formal descriptions and do not have the resources to construct an adequate *human* language for God, certainly not one which Aquinas would recognize as 'what all men call God'. Considerations of value also have to enter our judgments about what are good models of God and in what language we can describe him. There are no established religions that do not contain some elements of personality and human value within their model of God, and these cannot emerge from scientific explanation alone.

The important thing is to give an interpretation of how Aquinas conceives his so-called arguments. He concludes each with the phrase 'and this is what all men call God'. In other words, given a prior and implicit understanding of the concept 'God' that he believes to be universal, these evidences from the creation lead us to identify him with all the existences required to account for the world as we find it. The arguments are rather illuminative than deductive; *from* a preunderstanding of God *to* the world rather than the other way round. They are certainly not deductive, because there are crucial shifts of meaning in each. The first 'cause', the 'necessary' and 'perfect' being are explicitly said *not* to be exactly like the causes, perfections etc. found in the world. They are attributes described *analogically*.

What Aquinas meant exactly by this is not easy to determine. First he attempted a careful balance between the positive and negative features of an analogy: God is *like* a cause in being the first term of a series of what we normally call causes, but just because he is the first he is *unlike* all the rest. Sometimes Aquinas stresses the 'negative way', that God is necessarily and infinitely other than the creation (otherwise he would not be 'what all men call God'). His essence as he is in himself is unknowable and certainly cannot be expressed in human language. This is the *agnostic* element of Aquinas' theology. On the other hand, if this were all, how could we speak of God at all? Why use the word 'cause' if it is an utter equivocation on our normal usage? It cannot be an utter equivocation, but neither is it literally our normal usage; it is analogical.

As for what it means to be 'analogical', Aquinas has little more than negative characteristics to offer. There is always the danger of swinging from equivocation and hence agnosticism to *anthropomorphism* in trying to understand the meaning of human language when applied to God. Neverthless we must use human language or remain silent.

Wittgenstein's instincts were right when he made this point at the end of the *Tractatus Logico-Philosophicus*, but unlike Aquinas he then regarded language as simply univocal, as he had to if it were to be a fit instrument for 'logico-philosophy'. Later of course he rejected this prejudice, and in his theory of family resemblances made the first steps in modern times towards a theory of analogical use of language. It cannot be said that we have got much further than Aquinas or Wittgenstein in developing this, although it seems to me that an adequate theory of analogy is essential before we can begin to do justice to the theological use of language (Arbib & Hesse, 1986, chapters 8, 11; Hesse, 1988).

Meanwhile some light may be thrown by comparing models of God with models in physics. According to the moderate realist view, physical models also have positive and negative analogies with the observable world. Physics cannot claim to capture the essence of 'reality', and even if it could in principle do so, it certainly has not reached any such goal yet. But this does not imply that we are forbidden from speaking of a natural reality, whose essence we do not know, but which constrains the outcomes of our experiments and the modest local theories we use to represent them. The ontological belief that there is a constraining reality, and that we can even refer to it, does not depend on being able to represent it accurately, as we saw in the case of Priestley and phlogiston.

The way scientific models relate to the world is linguistically very similar to the way models of God relate to – what? Well, models of God do not describe his essence, and are subject to radical changes from place to place and in the course of religious history, just like scientific models. They do also relate to some things that are observable: to religious formalities, institutions and experiences, such as prayer, worship, ritual, community ethos, spiritual insight, and as Durkheim said, to social cohesion and survival. It is these that must act in any given community as the constraints upon the adequacy of its religious models. The facts and values inherent in such religious phenomena mean that there are good and bad models for God as there are for physics relative to its phenomena.

To leave it at that would be the equivalent of simple instrumentalism in science, and most religious believers want something more realistic in their models for God. They want to believe in some reality constraining all that variety of experiences which is interpreted by them in the 'theory-laden' language of religion: God's will for the individual's life, the communal search for meaning and goodness, the place of humankind in the Universe, and all the human responses to these in prayer, penitence, joyfulness and worship. To believe in and refer to God in these ways does not imply that we accurately represent him, any more than theories accurately represent the natural world.

The kinds of evidence involved in religion are of course different from those of natural science, and this is where the crucial epistemological difficulties of religion begin. We can give a moderately realistic account of scientific models because there are well-known and acceptable criteria of good and bad scientific theories and for making relative progress within limited domains of application. There are local pragmatic successes of prediction and control. To say this is to reject an undifferentiated pluralism about science, and also a strong social constructivism, although it does place the 'well-known and acceptable criteria' firmly within the social rather than the purely rational context. Well-known and acceptable to whom? To a western-based society, barely 300 years old, which chooses to adopt as a measure of objective knowledge what the seventeenth century called the 'experimental philosophy', with roughly Baconian criteria of success.

Past religious ages did not adopt this measure of objectivity. They had others, such as divine revelation in scriptures and in the individual soul, the authority of religious institutions, sometimes linked with state power, and certain a priori rational and ethical values of the mind. Our current difficulty about models of God is precisely that we have no socially acceptable standards of objectivity and truth that will accommodate such theological (or even ethical) talk. The models remain, as far as their epistemology is concerned, imaginative constructions inherited from tradition, or woven anew by individuals and groups who reject a purely naturalist vision of the world.

It is in such imaginative constructions that we should find the true import of the science-based models: they are metaphors for God, not metaphysics. Nevertheless metaphors and analogies are not to be devalued; they and not metaphysics are in the end the only way we can speak of God, and indeed also of natural reality. Metaphors do not exclude belief in God's existence, although they cannot totally represent him. Metaphors may be better or worse, and appropriate metaphors may even come from natural science itself, but the criteria of appropriateness must arise chiefly from domains of human life, history and society that transcend science. I hope I have indicated in relation to a few types of model at least the sorts of criteria these would be.

## Bibliography

M. A. Arbib & M. B. Hesse, 1986, *The Construction of Reality*, Cambridge University Press.

P. Davies, 1992, *The Mind of God*, Simon & Schuster, New York.

M. B. Hesse, 1988, 'The cognitive claims of metaphor', *Journal of Speculative Philosophy*, **2**, 1–16.

M. B. Hesse, 1992, 'Science beyond realism and relativism', *Cognitive Relativism and Social Science*, eds. D. Raven, L. van Vucht Tijssen & J. de Wolf, Transaction Publishers, New Brunswick NJ.

# Discussion

The following questions and answers have been selected from the discussion at the Symposium for the purpose of clarifying and supplementing the preceding texts. All authors were given the opportunity to reconsider and revise their answers. The names of those who put the questions have been added only when this makes the text more comprehensible.

## Questions to Gerard 't Hooft

*Question* (Barrow): From the last part of your lecture I conclude that you believe that the development of the Universe can basically be seen as a kind of discrete computational process and that we should attempt to derive the continuous laws of physics from that discrete process.

*Answer*: There are several indications that ultimately there is some sort of discrete structure of space–time, although you cannot really determine in what sense it is discrete. Mathematicians would tell you that continuous and discrete theories are related to each other by various sorts of theorems. In the end it may be that there are several equivalent ways of describing the same mathematical structure, a kind of complementarity. At present there exists no theory whatsoever that completely combines what we know about gravitation on the one hand and quantum mechanics on the other. String theory can at best be only a partial answer. The complete theory does not yet exist, because it seems that the combined requirements from the theory of gravitation and from quantum mechanics form together a symmetry pattern which is too difficult for us to understand. We have absolutely no model that obeys these symmetry requirements completely. This seduces us very much into thinking that if there is a solution at all it might well be a unique solution. And then you have a Theory of Everything.

*Question* (Davies): You showed us that in this Conway game you can watch very complicated patterns from a simple starting configuration just by letting the computer run for a long time. You said also that these patterns were 'complex'. It seems to me that you are using the term complex in exactly the other way around as John Barrow, because your patterns are of course algorithmically compressible to a very small set of instructions.

*Answer*: Of course, you are quite right. We see all these apparently complex patterns, like us sitting here, and we do not know how to compress these into a single law. But if they were compressible to a single unique law the complexity would become spurious, there remain only appearances. If you create a pattern within such a life game it could appear to be equally as complex as indeed the patterns we see in the Universe. Perhaps you should call these appearances of complexity.

*Question* (Dieks): How can you say something about a Theory of Everything by studying life games, because it seems to me you can have a lot of initial conditions compatible with the rules, so it does not comply with your second requirement.

*Answer*: Of course the boundary conditions are often not yet given in the life games. A complete Theory of Everything must include all details about the boundary conditions.

*Question* (McMullin): In the context of Theories of Everything and this Conway game of life you mentioned both the term law and the term theory. Could you help me to understand what for you the difference between the two terms is and specifically what you mean by 'theory'.

*Answer*: In a *model* you only ask which consequences follow from some arbitrarily chosen laws, without stating whether it corresponds to the real world. I would call a model a *theory* as soon as I believe that it has something to do with the real world and could even explain some really observed phenomena.

*Question*: The parapsychologists have tried to strengthen their case by referring to quantum mechanics. The EPR (Einstein–Podolsky–Rosen) paradox is taken to be a proof for the possibility of clairvoyance. (In this respect the Geneva conference with Arthur Koestler and the Cordoba conference 'Science et Conscience' with Costa de Beauregard come to mind.) So far the standard reply has been: there is no information transfer without energy transfer. Is that correct?

*Answer*: Yes.

*Question*: In your talk you seem to suggest that one can eventually realize the EPR paradox in Conway's game of life. Now, I don't know much about this topic, but I can imagine one could mimic some kind of EPR paradox 'life'. However, Conway's game of life is causal and the whole structure would eventually emerge from a well-defined set-up at the beginning.

*Answer*: I discussed with Bell the question whether, in the case of fluctuations in his vacuum, which are space-like correlated, his inequalities would still hold. And it turned out to be a very difficult question. You have the same situation in these life games in which you create an apparent chaos (which is compressible). I believe it is possible to construct an EPR-like experiment in such a kind of life game. I even tried it and found some mathematical schemes where it seems to work. But these schemes all show failures of some sort which prevent me from drawing definite conclusions. At present I am not convinced you will still find the Bell inequalities in these systems.

*Question*: Are scientists influenced by philosophy?

*Answer*: I *do* think that scientists *are* being led by certain philosophical concepts and attitudes. I think the work of a scientist contains two kinds of ingredients. (1) When he has a theory he must find ways to test it, devise experiments etc., and if he has theorems he should try to prove them. (2) He has to try to devise *new* theories, ask *new* questions, and judge the relative *value* of theories and questions to be asked.

Now in the first case philosophy, I think, is not so important, but in the second case it is. Let me give two examples: when in the beginning of this century quantum mechanics emerged as a theory, physicists were evidently forced to abandon certain realist points of view and replace them by a more empiricist attitude. But when, twenty years ago, quark theory was proposed, many physicists, particularly those who had witnessed the changes due to quantum mechanics, opposed the theory: quarks could *not* be observed as physical particles, therefore should not be viewed as a reality. This time it turned out that you had to have a 'realist' attitude. Nowadays quarks are considered to be exactly as real as electrons.

My conclusion from these events is that if your philosophy is not exactly right you might be fooled by faulty intuitions.

*Question*: What does 'truth' mean in science?

*Answer*: If you are an American or a British citizen you had better believe in the existence of some truth, because if you appear in court you have to swear to tell 'the truth, the whole truth and nothing but the truth'. For a scientist in my opinion truth is a one-dimensional concept: something may be true, an appearance, a false appearance or an outright lie. And science, in my opinion, is making progress, and in doing so it is continually replacing lies by truths, but not complete truths. For instance, science will never be able to reconstruct all the thoughts this audience is having right now. We don't know what they are, and we probably never will. Now it is quite correct that there could be more than one 'truths'. For instance, Newton did not need the gravitational field to solve his equations; nowadays we say that this field is 'real', we even believe there are gravitational ripples, waves, that propagate. You can describe Nature in terms of second order differential equations or by first order equations. Mathematicians know that then the number of 'real' variables doubles. You could call some of them 'hidden variables' (admittedly not a good choice of wording). This is just one example of a case where you can represent exactly the same phenomena in terms of different existing entities of which you can say that they 'exist'. They are 'true'.

## Questions to John Barrow

Question (Hesse): Do Theories of Everything give new predictions as well as values for physical constants? Einstein's universal field theories were criticized on grounds of being merely unifying devices of laws already known, without predictive power.

*Answer*: One should not imagine that a Theory of Everything determines necessarily the physical constants uniquely, because there could be fluctuations and quantum theory can produce only probabilistic results. But even if the theory is going to tell you all the constants it will not necessarily give a nice list of exact values comparable with laboratory results. The theory may for example just predict the values effective at very high temperature. To figure out what they will be at low temperature in your laboratory you need to know the early evolution of the Universe. If the theory has a large number of dimensions – some of them predict 26 – then all except the three space dimensions of our world must be fantastically small. And until we understand how they became so small, we can't trace the predictions from the early stages to the present.

*Question* (Drees): My question concerns the distinction made between 'Platonic' approaches (laws, symmetry, elementary particle physics) and 'Aristotelian' approaches (complex outcomes, broken symmetries, biology etc.). What place is given to the laws with respect to reality? Is there any sense in which laws have priority over outcomes? Would that be a priority of temporal nature, thus claiming that laws precede the physical universe? Or, also temporal, that the very early Universe was simple with the laws more directly 'visible' than in a complex phase. Or is it a priority of a non-temporal nature, so that 'precede' would not be an adequate term.

*Answer*: I think it is not possible to figure out the nature of all the outcomes of the laws from the laws alone. Likewise it is not possible to deduce all laws simply by studying the outcomes. If you have a sequence of events, which is incompressible, it means really that you cannot reduce them to some law. Equivalently there are statements of physics which are undecidable. The question for example whether an orbit of a Hamiltonian system is chaotic or not is undecidable. It is because of this interlinking of the laws and the outcomes – and people don't usually distinguish between them – that one has a conflict between empiricists and realists and this is a healthy thing. But there is enough room for both groups of philosophers to study different parts of science, actually without coming into contact.

*Question*: A main topic of your lecture has been whether the Universe is simple or complex. It seems we have a peculiar mixture of both. On the one hand we have simple substructures which are highly compressible into physical laws and on the other hand we have boundary conditions and symmetry breakings which seem to be very complex and which are not or hardly compressible. The boundary between them however seems vague and to some extent undecidable. How much of this may finally turn out to be illusory, as we have seen laws come and go or have only a restricted or approximate applicability?

*Answer*: We are not only thinking that the Universe is compressible, but in effecting this compression, our brains do all sorts of unusual things like throwing away information or finding compressions where maybe none exist. We need this process to survive the evolutionary process. If you see tigers between the trees when there are no tigers between the trees then your friends will just call you paranoid; but fail to see tigers between the trees when there are and you'll be dead. The evolutionary process selects for oversensitive pattern recognition capabilities. Clearly there is a limit to how much of this redundancy is worth building into an organism. Similarly, if we could record all the information on offer about our environment with our senses, right down to the scale of atomic levels, we would need unfavourably large facilities for information storage and processing.

### Questions to Ernan McMullin

Question: Does the supposedly greater explanatory power of causality arise from its anthropomorphic nature?

*Answer*: Not at all. It is true that we as children come to learn what it is to 'cause', to bring about, an effect through our own experiences in interacting with the world. But the reason why explanations specifying causes have greater explanatory power is quite simply because they tell us *why* the events in question happen (or happened). They answer the question: why? Simply situating an event as an instance of a larger regularity (as the positivist model of explanation suggested) is of itself not enough.

*Question*: Mechanics, as you said, is a rather special type of physical theory. On the other hand, for such sciences you mentioned, like chemistry or palaeontology, Shapere has argued that the other kinds of knowledge they gather is not theoretical knowledge, but rather knowledge aimed at some sort of systematization. Is your restriction not weakening realism too much?

*Answer*: I don't think so. First, we *do* have theories in such sciences as chemistry and palaeontology. Chemical theories postulate molecular and atomic level structures and processes that are causally sufficient to explain laboratory findings. Palaeontological theories reconstruct in a tentative way the sorts of beings that existed at some point in the distant past in order to explain in a causal way the traces of those beings we find in the present. These theories systematize our knowledge, of course; all theories do. But what is far more important is that they are *ampliative*; they enlarge our world. They tell us, with some degree of likelihood, of things that exist, or existed, in realms far out of reach of our senses.

Second, the reason why realism seems a problematic doctrine if one focuses (as all too many do) on the domain of mechanics only, is that explanation in mechanics (unlike chemistry and palaeontology) is only doubtfully causal. Newton used terms like 'force' and 'attractions' as though they specified causes, but would then add phrases like 'the cause of gravity I do not know', and would treat 'explanations' in terms of force effectively as descriptions in verbal disguise. The success of a theory such as Newton's cannot easily be interpreted in realist terms, simply because we don't know what is being claimed to exist. Is space–time structure in Einstein's General Theory of Relativity a descriptive account of the way bodies move when set in certain configurations, or is it the cause, the agent, bringing those motions about? The success of Einstein's theory does not carry with it a corresponding claim about real causal structure (as it would for a chemical theory) simply because there is no agreed ontological interpretation of the successful formalism.

*Question* (Hesse): I agree with you that the criteria for scientific realism consist of lists of requirements in local situations, and they are neither necessary nor sufficient. This is a type of 'moderate realism' that defuses the philosophical issue and replaces it by case-by-case considerations. Perhaps it's time to bury the philosophical issue. But what do you think about questions of 'convergence' on reality? Do you follow Kuhn on revolutionary theory-change?

*Answer*: You are quite right in suggesting that my approach deflates the problem of realism, and prevents it from becoming the sort of intractable metaphysical issue that many of its critics have managed to make it seem. Arthur Fine remarks that the difference between his 'natural ontological attitude' and the realism he objects to is that the realists, though they say the same things about the world as he does, bang the table while they are doing so. Well, you may have noticed that although I may wave my hands a bit, so far I haven't banged any tables!

The question of convergence is a particularly controversial one. I have argued for a progressive discovery (part discovery, part construction) of underlying causal structures. It is easy to document this sort of steady progress in the knowledge of specific structures in such fields as geology, chemistry, cell biology, astrophysics. (Remember, once again, that mechanics has to be treated as a *peculiar* science!) Even the most foot-dragging of empiricists would admit much of this progress, though he would dispute that it was the *explanatory* success of the correlative theories that brought it about. One symptom of this success is what I have called 'consilience', the way in which a good theory unifies domains previously thought of as disparate. So that there is at least this sort of constant convergence as theory progresses. Where the claim becomes controversial is where one is asked: how close to the end are we? Can we ever know we are at the end? Is there 'one true theory' somewhere at the horizon? We can tell how far we have come; we do not know how far there is to go. And I doubt whether there is a *unique* mind-independent ontological structure of the world towards which science necessarily converges.

In his Afterword to a newly-published Festschrift in his honour (*World Changes*, edited by Paul Horwich), Kuhn is still the resolute anti-realist, because he believes that his account of revolutionary paradigm-changes rules out realism. My response would, of course, be that to the extent that it does, there is something wrong with the account! More to the point, there is a problem of consistency here; in *The Essential Tension*, Kuhn insists that he does not challenge the *objectivity* of science. As I have noted in my paper, the argument he uses in support of this claim ought, as far as I can see, to lead him to accept some sort of realism also.

*Remarks* (Brümmer): (1) I agree with you that 'realism' is a highly ambiguous term. This has to do with the fact that the term 'real' is also ambiguous, and is often used to deny something. Its meaning varies in accordance with what it is meant to deny. Thus 'real' can mean: non-bogus, non-fictitious, non-illusory, etc. In all these senses something could be claimed to pertain to the way the world is, as opposed to the way we imagine it, we visualize or we think it, etc., to be. A rejection of realism could be taken either in a universal sense as the denial that these distinctions make sense. This would amount to a rather implausible kind of phenomenalism. Or it could be taken in a limited sense to mean that certain claims (in science, religion or ordinary life) do not pertain to the way the world is but rather to our illusions or fictions about the world. Such limited claims are meaningful since they are specific applications of the real/non-real distinction and not rejection of it.

(2) I agree with your remark that 'truth' is not a criterion for reality. On the contrary 'truth' (like 'goodness' or 'beauty') is a term of appraisal which is not a criterion but rather requires criteria for its application. In this sense Richard Hare's distinction between 'meaning' and 'criteria' (with reference to 'goodness') also applies to 'truth'. Thus the correspondence theory of truth is a theory about the *meaning* of a factual proposition: it claims to state how the world is and not merely to our illusions or fantasies or cognitions about it (in this way the concept of 'truth' presupposes the real/non-real distinction). The coherence theory and the pragmatic theory of truth are theories about the *criteria* for truth: a factual proposition is true if it is entailed by a theory or a frame of reference, etc., and the theory or frame of reference, etc., is true if it is in some way adequate for coping with our experience of the world.

*Reaction* (McMullin): Very well put. Correspondence, as you say, bears on the *meaning* of a truth-claim, whereas coherence and pragmatic factors bear on its *acceptability*. The custom of using the same term 'theory of truth', in discussing all three, as though they were competing alternatives, is grossly misleading. To illustrate your point about the use of 'real' as a contrast-term with 'illusion', let me recall

Ian Hacking's version of realism, which he calls 'experimental' realism. Like Kuhn, Hacking believes that theory is too unstable a support for claims about the real. But he argues that experimental intervention can convince us that a particular entity our instruments appear to register is real rather than an artefact of the instrument. The critic who has to be convinced in this case is not an anti-realist but simply someone who wonders whether the entity apparently observed by the particular instrument may not be an illusion.

*Question* ('t Hooft; also to van Fraassen): In a given area of research I can imagine three different situations when it comes to producing a theory that claims to be able to account for the 'observations'.

(1) There could be one theory that describes just one 'truth' such as: 200 million years ago there was a supercontinent. I suspect that neither the realist nor the empiricist will have any difficulty in accepting the theory and working with it.
(2) There could be two or more theories, all describing some 'truth', but the different 'truths' are incompatible. This situation might actually occur for the inhabitants of a cellular automaton, who could discover different sets of laws, different 'Theories of Everything', that predict exactly the same phenomena.
(3) There could be a theory that, when interpreted, allows for no single 'truth' at all. I am referring to the situation with quantum mechanics, where either the positions, or the momenta of the particles are 'real', but not both.

I would like to know how the realist and the empiricist would react in the three situations and to what extent they would be satisfied with such theoretical accounts.

*Answer* (McMullin): Taking your three situations one at a time. As a realist I have no difficulty with the supercontinent of some hundreds of millions of years ago. But the strict empiricist ought, I think, to balk at it. Geologists defend this theory on the grounds of its explanatory success; they believe such a continent existed because the theory is a highly successful one. The retroductive inference from observed effect to (in this case) unobservable cause ought, however, to be a problem for the empiricist. I have always wondered how so conscientious an empiricist as Bas van Fraassen can take the theory of evolution seriously as an account of what *really* happened in the distant past, when there were no empiricists around to observe!

Regarding the second situation, I do not think there can be two *genuinely* different theories (that is, theories involving incompatible ontologies) that account for all possible observations equally well. (Here I am disagreeing with Quine, among others.) There may be two theories that account equally well for all observations up until now, but one would in that case expect them to diverge in the future. As Kepler said when discussing this same issue: falsity will ultimately show itself. Jim Cushing has recently argued that the Copenhagen and Bohm interpretations of quantum theory constitute essentially two different theories (since their ontologies are quite different) whose predictive consequences are, in principle, the same in all possible circumstances. I am perplexed by this, but would remark that, even if that be so, there might in principle be another means of deciding between the theories based on the presumably different explanatory virtues of the two theories.

As to the third situation, I don't think there can be a theory that allows for no simple truth at all. The situation in quantum mechanics where the positions and momenta of elementary particles are in some sense (according to the Copenhagen interpretation, at least) indeterminate does not affect the truth of the theory. Nor does it prevent there being truths about the positions and momenta. The theory simply sets limits on which pairs of truths can be co-affirmed. The *truth* about the position of a particle might well be that it is indeterminate. Truth does not require determinacy.

### Questions to Bas van Fraassen

*Question* (Drees): When you presented science as an 'open text' you took the multiplicity of interpretations of quantum theories as your case. Professor McMullin placed physics in a category by itself and said that the historical type of natural sciences, such as plate tectonics, are much stronger cases for realism. Could you present your position also by taking a case from geology, such as plate tectonics, or would that differ in relevant aspects from the case of quantum physics.

*Answer*: A theory will admit multiple interpretations if it leaves any questions open. If they are empirical questions, however, the answers will amount to additions which change or increase the theory's empirical predictions, so more than 'merely' or 'purely' interpretation is at stake. I know very little about geology, and I do not know whether the open questions there are all empirical, or whether, there too, there may be several ways of understanding a theory which do not affect its empirical content.

The connection between realism and interpretation has of course to do with truth. The scientific realist presumably argues that quantum mechanics, for example, may be true under one interpretation and false under another one. That will depend on whether there really are many worlds, or whether presence of consciousness affects physical processes, and so forth. Then, if the aim of science is to give us true stories about the world, that aim will include arriving at an interpretation of quantum mechanics under which it is true. Unfortunately – and this is what may make the case of quantum mechanics so much more interesting than that of geology – there cannot be an experiment which confirms quantum mechanics but tells for or against, say, Bohm's interpretation as opposed to the Many Worlds Interpretation. For these interpretations are designed not to affect the empirical predictions in any way. Therefore, if the scientific realist is to argue along the above lines, s/he has to come up with ideas about how non-experimental 'evidence' can give reasons for belief, or for greater likelihood of truth.

Professor McMullin confronts this problem head-on in his paper when he discusses what he calls 'complementary virtues' of theories, such as coherence. He uses the Copernican debate to show how important a role these play in scientific theory evaluation and theory choice. According to him this makes little sense from an empiricist perspective. But I would say it makes a different sense: these complementary virtues have a pragmatic character, and play a role as good reasons for acceptance (including commitment to bases for further research) but not for belief. They do not make the theory any more likely to be true.

It actually makes me rather happy to hear Professor McMullin say that study of physics tends to lead people to anti-realism in the philosophy of science, while study of biology or geology does not. If it were the other way around, I'm sure realists would urge us to look at physics, so as not to be deceived by surface appearance.

*Question*: Should philosophical views on the scientific enterprise affect the way in which a scientist works?

*Answer*: If everything is going well it will make no difference. The realism–empiricism debate is about how to view science, not about how you, as a scientist, should act. When scientists face difficulties that seem to them to be very fundamental, they naturally and I think rightly turn to philosophy. For it is at such points that ways of looking at things, ways of understanding the information you have, rather than getting new information, can take on crucial importance. Different possible interpretations of an old theory may harbour different suggestions for developing a radically different rival theory, that could do better than the old one.

*Question*: Since the empiricist view on science seems to exclude a physical world-view, does it also exclude a Theory of Everything (which seems to imply a world-view)?

*Answer*: It is sometimes argued that on empiricist grounds, it would be just as good if the sciences were totally splintered and disconnected. After all, if we don't want to find the One True Story of the World, why would we be interested in arriving at a single story at all? But there are good practical reasons, from within the sciences themselves, to try for unification. These reasons appear every time we focus on some area of overlap. If medicine and physics provide only mutually incompatible or totally disconnected stories, how do you develop space medicine for astronauts? The drive toward unification in physics is a very natural development, which it does not take realism to explain. Besides the practical reasons present everywhere in science (which establish the unquestioned background assumption that unification is a good to be pursued in general) there is also something else at work however. That is the fascination with symmetry arguments, and the hope that by putting in enough symmetry constraints, the theory being developed will be unique. But the result could also be very shallow, if this can be done only at a very high level of abstraction.

*Question*: I would like to have some clarification as to what you said about the role of mathematics in science. As you said, according to the Anglo-Saxon tradition, mathematics enriches language. According to you it does more; it has a function for constructing models. In what sense is this different from the role of language, which may also serve as a tool for constructing many kinds of models?

*Answer*: It is a bit disconcerting to think, of course, that this 'Anglo-Saxon' view derives largely from such philosophers as Frege, Wittgenstein, and Carnap, so the term has only rather recently become appropriate! But the view does contrast with, for example, that of the Intuitionists, who gave language no function in mathematics except for the communication of results of construction in imagination or other mental mathematical activity. In Platonist philosophies of mathematics, the activity is not conceived as analogous to artistic construction of sculptures or paintings, but to the description or display of abstract *objets trouvés*. I suppose to a Platonist, if mathematical is like artistic activity it must be like that of Marcel Duchamps with the urinal. One of my great regrets in life is that I do not have a philosophy of mathematics; I will just assume that any adequate such philosophy will imply that what we do when we use mathematics is all right.

*Question*: You say that theories are only rivals if they differ empirically, but this isn't always easy to see. Theories are never completely worked out, it takes a lot of commitment and effort to work a theory out. The rather easy idea you have in empiricism may not be very practical.

*Answer*: I agree. In the midst of scientific toil and strife, the conditions for philosophical discussion are usually not met, and the theories being debated are often not yet in the sort of shape that allows for clear philosophical questions. The same is true for law, politics, war, . . . and any theoretical discussion thereof. It is also true that there are many things we can never find out without sticking our necks out. I've found this out in rock-climbing: often you cannot discover whether something really is a foothold without committing yourself to it. Then if it isn't, you fall . . .

*Question* (Muller): I was wondering what the difference is between your credo 'make empirically adequate models', and Popper's, by now 60 years old credo 'make empirically refutable theories'?

*Answer*: Popper made it a criterion for a scientific theory that it is to be refutable. That was really too strong, as he and his followers found out, exactly because observation reports do not carry the absolute authority that would be needed for refutation in a strict sense. So they weakened the criterion in various ways. To be scientific requires that you are willing and ready to give up your theories if they meet with a certain (unspecified) amount of contrary evidence. It also requires that the theories you advance are audacious, so that it is quite possible for them to meet with such ill fortune.

While this is most easily understood as something like a prescribed scientific 'ethic', I do think it is closely connected with the issue of empirical adequacy. That is what I regard as the bottom-line scientific criterion of success. But just like the stronger criterion of truth overall (which I do not so regard), it is very easily met if you give up all audacity. You could just speak in tautologies; then your statements will be sure to be empirically adequate and even true. That is in fact what people seem to do when they feel emotionally troubled: it is comforting to think that *whatever will be, will be*, and so on. But this also shows that meeting the bottom-line criterion is not enough to make a thing be of value. So empirical strength (informativeness) is the very next criterion.

*Question* (Muller): You have conceded that it is rational to use abduction, that it is rational to believe that electrons exist, and that it is rational to be a realist. Still you maintain that realism is not required in the scientific enterprise. I would like to focus on that point.

The normative part of constructive empiricism seems to consist of no more than the slogan 'make empirically adequate models'. I want to ask something about the descriptive part. In one of the essays we had to read for the symposium Hacking tells the story of an experimentalist who wakes up at night and thinks: Oh dear, if invisible parts of dust fall upon the surface of the experiment they could get magnetized too and interfere with the effect under investigation. The next morning he runs into the laboratory and treats the surface of the experiment with anti-dust spray for a month. Many more examples could be added to show that it seems essential to be a realist in order to be a good experimentalist.

Concerning theoretical science, the existence predictions, like Neptune's, the positron, the neutrino, the omega-particle, but also missing-link predictions in the theory of evolution, seem to be very difficult to understand without the realism of their inventors. If the constructive empiricist description of science does not include realism, is it not simply empirically inadequate?

*Answer*: I have conceded that it is rational to use abduction only in a very limited sense! If on a certain occasion someone makes up his mind to believe the very conclusion that abduction would lead to, that is not irrational. Of course, it is also the very conclusion he would be led to by the rule to infer just anything from anything – and by many other rules as well. To adopt abduction as your *rule* for how to change your mind in response to evidence and new experience I do not consider rational. The reason is, as I argued in *Laws and Symmetry*, that this leads to incoherence in one's opinion.

I do agree that it is not irrational to believe that electrons exist, or to be a scientific realist – and I think that everything is rational if it is not irrational. But I do not think that science requires any of this. You see, I could add a little to the story of the scientist who had the nocturnal experience Hacking described. As it happens, this scientist does not believe in the reality of magnetic forces. But the experiment he was doing was designed *on the basis* of a certain theory; that is exactly why the outcome of the experiment would be relevant to the question of empirical adequacy of that theory. This theory says that there are magnetic forces, and so as long as he ignored that part of the theory, his design was defective.

To this we should add that he knew very well, as do scientists, detectives, biographers, and many others, that people do their best work if they completely immerse themselves in their subject matter, which involves a great deal of thinking 'under supposition'.

To sum up: you cannot infer from a scientist's behaviour, or even from the way the scientist thinks when s/he is engaged in scientific work, whether or not s/he is a scientific realist. The question whether or not s/he believes that the unobservable entities exist is independent from his or her philosophical point of view. (A constructive empiricist could rationally make up his mind to believe more than he is rationally required to believe.) But the same point applies there as well: not only can you not infer the person's philosophical position from action and thought while actively engaged in a scientific project, neither can you infer from it what this person believes to exist.

*Question* ('t Hooft; see questions to McMullin).

*Answer*: I certainly agree that the second situation can occur. Your example of the cellular automaton illustrated this possibility very effectively. In my terminology, there can in principle be two theories which are incompatible but empirically adequate. (It surprises me that Professor McMullin does not appear to agree with this; that is one way in which he appears to disagree with other realists as well.) From the empiricist point of view, such a pair of theories would each serve the aim of science equally well, and there could be no good reason to think that either of them is more likely to be true.

There are several factors that tend to confuse this situation, but I submit that the distinction between belief and acceptance removes the confusion. To accept a theory (leaving qualifications aside) is to believe not only that it is empirically adequate, but to have also a certain positive non-epistemic attitude toward it, such as a commitment to make it the basis for research, to attempt to integrate other parts of science with it, to draw on it for explanation, and so forth. This additional endorsement may well involve some beliefs that do not relate directly to the question of the theory's truth – for instance that this sort of research is likely to be vindicated in the short run (such as one's own lifetime). It may also be due in part to the belief that none of the rival theories available at the moment will be integrable with a certain other part of science but that this one will be the easiest to modify appropriately. The point here is of course that to modify a theory is to *reject* it and replace it with another one, a near-variant but still a rival. This is the sort of victory in defeat that one sees very often in science.

To the third situation I have a very different reaction. I am not sure that the example from quantum mechanics that you give brings out everything that you have in mind here. On the orthodox interpretation, a particle cannot have both a definite position and a definite momentum. But it is not as if we have a choice: the state of the particle decides unequivocally which, if either, it has. On Bohm's interpretation, on the other hand, the particle does at all times have a definite position and momentum.

Perhaps the most telling subcase of the third situation you describe would be that of a theory that admits of various interpretations, each of which is in some way unacceptable to you. Of each you would then say: 'the world could not be that way!' It is easy enough to sympathize with that in the case of quantum mechanics – each interpretation makes the world look 'reasonable' in some way, but very weird in other ways. However, it seems to me that such judgments – about how the world could be, what is reasonable, intelligible, or unacceptably weird – are strongly culturally and historically conditioned. (At least I think they are when they go beyond questions of logical tenability.) As we look through the history of science we see striking differences in such judgments.

*Question* (Barrow): There is the possibility of thinking of the world as a computational, discrete process. This would rather naturally invite the constructivist view on mathematics as it is employed in physics. This would have consequences for what you might regard as being true. What is your reaction on this and are you happy with the excluded middle?

*Answer*: I am both happy and unhappy with the excluded middle. In a symposium which the logician A. N. Prior attended shortly before his death, I heard him say he didn't know what the fuss over Intuitionism was all about: either the law of excluded middle holds or it doesn't! Unfortunately for myself, I can't dismiss the issue so easily. I agree with the Intuitionists that the right way to make sense of mathematics is by focusing on mathematical activity – not by trying to construe the theories they produce as parts of a true story about some abstract reality. But I do not have a philosophy of mathematics; I cannot back up this conviction with a well worked out view.

The results of Feferman show that the mathematics needed by physics is a very modest part of Cantor's paradise, and there is much other evidence that physics might in principle be so developed as to draw on constructive mathematics alone. Well, suppose it is so. Physics so developed could also be written within a 'classical' mathematical framework. The two versions would not say the same thing, but their differences might be of no fundamental importance to physics. I am not at all sure however that these two options exist for physics as it actually is just now.

*Question* (McMullin): You claim that you are not an instrumentalist because (unlike the instrumentalist) you allow that theory does have the characteristic of being either true or false. But you are not a realist either because you hold we can never know which, and anyway this is none of the scientist's business. But you have just said that you would allow the claim 'electrons exist' as a 'rational *interpretation*' of quantum theory. It seems then that you do agree up to a point with the realist; realists too would call a particular realist claim (such as 'electrons exist') a 'rational interpretation' of the relevant theory. They would add on occasion, of course, that it is the *more* rational interpretation, which you would presumably question. Still, it does seem to make you a 'softer' sort of empiricist after all!

*Answer*: I do not want to identify scientific realism and empiricism as epistemological positions. What science is, is one question, and what we should believe is another one. In epistemology I do adopt a very libertarian position, and if that makes me a soft empiricist (or even, as you at one time graciously suggested, an honorary realist), I'll be happy with that. The only problem with libertarians is that so many of them think that extremism in the defence of liberty is no vice, and I'd prefer not to think like that.

But as our papers showed, there are still a lot of differences between us. Some of them are certainly in epistemology. One that occurs to me is that you seem convinced that Ockham and Berkeley *needed* theological premises to escape scepticism. (Perhaps they thought so as well; that is another matter.) This clearly has to do with your views on explanation: that if certain why-questions cannot be answered we are inevitably doomed to scepticism. But I do not see it that way. A second difference between us that also has to do both with belief and explanation is the status of abduction (what I call inference to the best explanation); but I have commented on that in my reply to Muller's second question.

*Question* (Brümmer, also to Feyerabend): Paul Feyerabend says that different people have different views and Bas van Fraassen said that this is a good thing. In brief, I would like to ask them: Do you consider yourself a realist, and if so, in what sense?

*Answer* (van Fraassen): I am an anti-realist. A realist thinks that our theories give us insight into the hidden structure behind the phenomena, into the causes and real reasons why things happen. These ideas about causality and the like are subject to so many difficulties themselves that I reject them as valid clues to the understanding of science. I reject realism as an unwarranted and unhelpful ingression of metaphysics which adds nothing to science. Most of all I reject the idea that realism (of that sort) is already implicit in the way we think, that it shows up the moment we draw on science to build bridges and make predictions.

I should perhaps have prefaced this answer by saying that 'realism' is an accordion term, on which we can play many tunes. But I hope I took the question in the sense in which you intended it. In view of your own discussion of language games, I should perhaps also add something about reductionism. I do not at all think that the language of science, which involves highly theoretical terms, is reducible to pre-theoretical or non-theoretical language about the observable phenomena alone. Nor do I think that this scientific language is dispensable to us. But to see how little is implied thereby, we have the example of eliminative materialism. This quite (though not entirely) well worked-out position was initiated by Wilfrid Sellars, elaborated in Richard Rorty's early work, and given great visibility in Paul Churchland's *Scientific Realism and the Plasticity of Mind.* It entails that our psychological language about persons and intentions and the like is not reducible to purely physical discourse, and that yet the entire truth about the world can in principle be expressed in physical discourse alone. While Churchland has argued for the piecemeal replaceability of all psychological language, Sellars envisioned such replaceability only in some ideal limit, leaving psychological discourse indispensable to us in any finite future. From this example we can see therefore – by analogy – that there is a great logical gap between language and reality, which will accommodate many versions of anti-realism.

*Answer* (Feyerabend): I consider myself neither a realist nor an anti-realist. I think introducing something like a form of realism makes sense, not in interpreting the temporary result of some complex scientific development, but the trend of the development itself. Your aim might be to find true theories. You try to reach this aim and you distinguish between what you want to find (which you might call reality) and what you have. Your realism might lead you to a new and different standpoint, for instance when you learn that it is the earth that moves and not the sun that rises and sets.

*Question*: Is it astonishing that our mathematics is applicable to the world?

*Answer*: There are many senses in which it is not astonishing but absolutely trivial, tautological, that mathematics applies to the world. The one substantive question I detect as part of this issue is this: is it not astonishing that we humans have so much mathematical ability? Of course it is not astonishing that we are better at maths than are snails or even chickens. But there still seems to be a basis for surprise, in that it seems that this ability would not have had any survival value in the early history of the species.

But evolutionary theory certainly does not imply that an inheritable feature can become prevalent only if it has survival value, or adds to fitness, on its own account (so to speak). It may be selected by Nature, without being selected *for*. One reason for this is that one trait can 'ride piggyback' on another trait that is being selected for. If traits come in clusters, the selection can only be between those clusters.

But aside from that, it is also not at all clear that the ability in question is not identical with another ability that did give the species an advantage – traits are not easy even to disentangle logically, let alone physically. As an instructive comparison I would here point to Stephen Jay Gould's essay on the evolution of the eye, with its memorable question: how could the first five per cent of an eye have given the organism an evolutionary advantage?

## Questions to Paul Feyerabend

*Question*: In *Three Dialogues on Knowledge* you formulate an argument to the effect that we don't know what success alternative sciences (like alternative medicine) could have for they are not the ones we have practised. What criteria do you have to judge those alternatives, and what criteria do you have to rule out the horrible uses of alternative methods. I am thinking of the racial theories in the Nazi period for example.

*Answer*: It is not up to me to judge the success of medical procedures; it is up to those who are using them or who consider using them in the future. Some central African tribes objected to being X-rayed. The function of X-rays had been explained to them but they just did not want their insides to be exposed to strangers. In such cases, I would say, the physician must find alternative and less intrusive methods. Secondly, it is an illusion to think that criteria for judging the outcome of novel procedures can be determined in advance. Novel procedures bring about novel situations in which existing criteria may become inapplicable. How do we determine that there is an object at a certain place having certain properties? Well, we look. That does not help us with shy birds or with wild animals whose behaviour will be drastically changed by the presence of an observer. What about the dark matter of the Universe? To start with it was a theoretical construct; it could not be seen. Naturally astronomers wanted more direct evidence. Then some astronomers dropped the 'prejudice' that whatever exists in the Universe must be visible. From now on it was the convergence of different theoretical results that served as a (temporary) criterion. And so on. What about the racial theories of the Nazi period? That is a very interesting question, especially today when some scientists think that social problems such as homelessness are, basically, genetic problems. And racial theories *could* be true, according to the best scientific criteria of the time. There is no reason why scientific results should agree with our ethical convictions. What should we do in such a situation? Well, here I have no problem at all. For me humanitarianism overrules scientific results liable to have inhumane consequences. After all, science is not sacrosanct. But love and compassion are.

*Question* (Hesse): You mentioned the changing 'social weight' of questions from Parmenides to Democritus: should we not therefore consider the social contexts of criteria for truth and objectivity? The seventeenth-century dilemma was how to acquire reliable knowledge in the midst of religious and political conflicts, when there were no decisive criteria coming from theology or metaphysics. Francis Bacon, Descartes and Boyle effectively proposed to define truth and objectivity in natural philosophy by the 'experimental method', which abstracted from all other questions of value. In the tradition of Bacon pragmatic success became most important. Whenever we talk about the correctness or acceptability of our world-view we are forced to talk somehow in terms of this new scientific understanding. We have nothing else we can put forward as clearly as the scientific institution is able to do. I think the scientific institution is a social phenomenon, that is demarcated from other institutions in society by its structure. It is world-wide and has roots in many different cultures. There is education, criteria for getting papers published, prizes and fellowships. And so there is a consensus among a community. (Admittedly not the entire population.) What is your view on this?

*Answer*: This is a very complex matter and the answer is not easy. So, let me make a few scattered remarks. Firstly, scientific understanding and scientific institutions. You speak as if they were uniform things. But they are not. In my seminar talk I tried to show that the phenomenon 'science' contains different and conflicting approaches. Secondly, pragmatic success does not always determine the actions of scientists. Despite numerous failures the problem of the stability of the planetary system continued to be pursued on a Newtonian basis. About thirty years ago engineering schools and businesses all over the United States (and also in Great Britain) replaced practitioners by theoreticians and a top-down procedure with on-site inspection and planning. There were numerous failures, the Challenger failure among them. (Read Feynman's report in his *What do YOU Care what Other People Think*?) Yet there have not been any decisive changes back to a more practical approach. Thirdly, the Catholic Church, during its heyday, was much more powerful than science ever was or, I hope, ever will be. A person without any religious affiliation was an unperson who could not be any citizen anywhere. And the Catholic Church was (and to some extent still is) a universal institution. Missionaries travelled around and converted foreign nations and tribes to their point of view, frequently by argument, not at all like scientists. It was in this situation, already softened by the Lutheran heresy, that modern scientists started looking for *and found* a new way. Are we supposed to be less imaginative? Besides, fourthly, new ways are slowly arising, at the boundaries of Western civilization and the so-called Third World. Most of the agents of development are scientists, or medical doctors. For purely pragmatic reasons they started mixing their science with local beliefs and items of information. 'We have nothing else we can put forward as clearly as the scientific institution is able to do' is therefore incorrect on many counts. There is nothing like 'the' scientific institution; what scientists 'put forward' is often unclear and full of conflict unless the conflict has been suppressed for PR reasons; and there are lots of things that are being 'put forward'.

*Question*: My question is this: when we are comparing world-views, is there something particular about the scientific one, historically speaking? Because rationality became associated to this particular institution of experimental philosophy.

*Answer*: Of course there is something particular about the scientific world-view, historically speaking. There was also something very particular about Nazism. The question is: does the particularity of science make it better than any other world-view? You say it is rational. Are you sure, or are you simply falling for scientific PR? Many moves within the sciences are 'irrational' in the sense that they conflict with popular versions of 'rationality' (such as falsificationism). Besides, why choose rationality as a conveyor of praise? Love, I would think, or compassion, is at least as important as rationality but it is excluded from the sciences. Which means that the sciences are lacking in important respects. Science cannot stop the killing in former Yugoslavia which would be more important than all the knowledge one might accumulate about the human genome. Let's keep a sense of perspective!

*Question*: Do you consider it to be foolish to demand empirical adequacy for science? And what about the demand for explanations?

*Answer*: It all depends on the situation. Plato suggested that astronomers should start with theoretical models 'and leave the heavens alone', as he expressed it himself. That was a sound intuition. Astronomical data are distorted in many ways. Giving a modern account we can say that there is precession, there are the mutual perturbations of the planets, refraction changes the path of light, our own position on the globe is rather idiosyncratic, the instruments are faulty – and so on. Starting with simple models we can isolate these effects and come to better results. This was the reason why Einstein said 'let the theory tell you what observations are and what they mean' and this was the way in which Heisenberg, following Einstein, was led towards the uncertainty relations (Einstein's theory of Brownian motion and his 1905 paper on light proceed in a similar way). Lack of empirical adequacy often results from a shoddy experience and only a non-empirical procedure can detect this. Explanation? Well, some scientists are interested in it, others are not. As always, notions that are loved by philosophers are used by scientists in a rather opportunistic way – they may accept them on one occasion, reject them on another, and they may not give a damn on a third.

*Question* (also to van Fraassen and McMullin): It has been proposed that philosophy of science can be seen as an empirical science of science. Do you think such an approach is possible?

And to Professor Feyerabend: if you think that this is possible, would your 'anything goes' theory not be empty as a science of science?

*Answer* (Feyerabend): An empirical science of science which pays attention to what scientists are doing from one moment to the next and which is not led astray by research papers or philosophical romances will come to the result that anything goes.

*Answer* (van Fraassen): To the logical positivist, philosophy was meant to be the logic of science. This idea is not bad, but their conception of it was too narrow. Lately there is a cognitive turn. To those people philosophy is investigating science in a scientific way. To me that is the worst kind of pseudo-science. They are doing an anecdotal study and draw far-reaching conclusions from that.

*Answer* (McMullin): I agree with van Fraassen. Next to empirical adequacy as a goal in scientific enterprise, I set causal understanding. Science of science is seductive, as we learned from Lakatos. You can trace back one part of science in many different ways.

*Question*: If I were to ask myself the question 'is there one Christian religion?' I could answer 'no' by arguing that there are sects. So if I want to argue that there is not one scientific world-view, I can point to many differences. But this argument is inconclusive, because there could be, and I think there are, very many things which scientists have in common. Could you give your comment on this?

*Answer*: Have you made a statistical study? Do you have the needed demographic evidence? I don't think you do. Neither have I but the anecdotal evidence I have found seems to point in a different direction as I have tried to argue in my seminar paper. There may be some widespread *slogans* such as 'we have to pay attention to experience' but even these slogans can't be found everywhere (they are rare in macro-economics, for example) and different people mean vastly different things by them (the 'experience' of high energy physicists has little in common with the experience of archaeologists, sociologists or demographers). And why are you so intent on similarit-ies? Isn't an enterprise that contains many approaches more interesting and more likely to have useful results than a science whose essence can be exhausted in three lines?

*Question*: In one of your books we had to read for the symposium, *Farewell to Reason*, you show a lot of sympathy for the medicine practices other than Western medicine. I agree with you that, in other cultures too, useful knowledge has been acquired during the past eons; it would be silly to deny *this*. But what we witness today is how Western medicine conquers the Third World, where everywhere health organiz-ations of purely Western stance are introduced. Would you say that this is just one form of unjustifiable post-colonial imperialism, or that this is because all other cultures are seeing that Western medicine is objectively better, in that by and large it heals more people and saves more lives?

*Answer*: You say that 'everywhere health organizations of purely Western stance are introduced'. It looks like it – but the micro-account is different. Large sections of the medical profession now turn from cure to prevention realizing that the environment and cul-tural factors play a large role in the latter. Medical workers in the so-called Third World have started combining their own information with the knowledge of the locals. It is true that Western medicines almost succeeded in eliminating certain types of infectious diseases and worms – but the diseases change, often as a direct result of aggressive medication and they can stage a comeback. Tuberculosis is an example, in Europe and in the United States. The manner of comeback depends on the habits of the populations which means again that there has to be collaboration, not imposition.

*Question*: Are scientists influenced by philosophy? Should they be?

*Answer*: Are scientists influenced by philosophy? Yes. Some of them were also influenced by alchemy and theology (Newton and Kepler are great examples) and some modern scientists regret that such influences are lacking today (example: Wolfgang Pauli, C. G. Jung, David Bohm and others). Should scientists be influenced by philosophy? Well, that is up to them; besides, I would hope that some philosophies would never even come near to the sciences.

*Question*: Are science and religion essentially different subjects?

*Answer*: Theology was the first discipline to claim scientific status and it was indeed more precise than the science of its time. In St Thomas both theology and science are subsumed under the conceptual system of the scientist Aristotle. Also theologians started the habit of settling problems by conferences – they called them councils. The rules of debate in the councils were much more strict than rules of debate at conferences have become today, though some conferences (I am thinking of certain conferences in connections with the claims of low temperature fusion) are rather carefully choreographed. There are of course differences but in the absence of detailed historical evidence I am reluctant to say that the distance between extreme positions in the sciences is always smaller than the distance between analogous ideas in science and theology.

*Question*: What does it mean to be a realist?

*Answer*: Hilary Putnam wrote a book *The Many Faces of Realism* – and he is right: 'realism' is a word that covers many things. Let me emphasize one aspect which I think is not unimportant. It is quite possible that a person possesses a certain amount of knowledge and, together with it, a certain view of the world and that the knowledge and the view are in conflict. Then, if the world-view is held with conviction the person may say: it seems that *A* – but in reality, *B*. Those who believed in inexorable laws underlying all that is had this position. They admitted that there were many exceptions but added that, having found the true laws, they would be able to account for them in terms of the laws. St Augustine had a similar view about miracles: many events seem miraculous, because they are unusual, or because they seem to conflict with past experience. But this miraculousness is simply a reflection of our ignorance, i.e. of the distance between appearance (what we know) and reality (how things are). Certain versions of realism are paradoxical; on the one hand they assert a gulf between human experience and 'reality' – on the other hand they use human experience to explore reality.

### Questions to Paul Davies

*Question* (McMullin): You want to talk about clever laws explaining how something came from nothing. First of all: the vacuum of course is not nothing. And more important: I can see space as finite, but unbounded, but I cannot see time in that way.

*Answer*: There are things which are in some sense logically prior as opposed to temporally prior: we like to think of the laws of physics as something existing in an eternal transcendent and timeless sense. But the idea that there is anything there before the origin of the Universe is swept away by thc notion of time and space coming into being with the origin of the Universe.

*Question*: How does the tendency to self-organization in the laws of physics combine with the second law of thermodynamics, in a closed system like the Universe?

*Answer*: According to the second law of thermodynamics, the whole universe must be running down towards heat death. Yet it appears to havc thc tcndcncy to bccomc more and more organized with time. There is no incompatibility between these things because the Universe starts out with a sort of stored negative entropy, which is used to pay for self-organizing processes. So you have to make a distinction between order and organization. In our universe organization progresses (advancing complexity), but the second law isn't being violated. The entropy is still rising. Whether the entropy or the organization will win, we don't actually know, because that's model-dependent.

*Question*: It seems to me that quantum mechanical indeterminism is a requirement for the spontaneous generation of the Universe. You cannot explain spontaneous generation within a theory about cellular automaton which is subjected to deterministic rules. Is that correct?

*Answer*: That is not correct. A big enough cellular automaton with random input would self-organize and generate shapes that could be considered to be living and intelligent. But I don't think we really explain anything when we say that the order we see has arisen by chance out of chaos. I feel a better route to go is the quantum indeterminism that allows a universe to come into being with a set of laws which already exist in some transcendental eternal sense.

*Question*: According to you, the case that the message 'Made by God' is all over Nature, would be a clear indication of the existence of a God. But would the message 'Not made by God' also do?

*Answer*: It doesn't matter what the message is, it hints at intelligence. Everyone has to make their judgement in which state they would be prepared to accept evidence for God as such. But you can of course always refuse to do that, even if the evidence is very persuasive. For, as we all know, there isn't even a way to be sure that you exist.

*Question*: In *The Cosmic Blueprint* you state that every level of complexity has its own laws. Insight in the higher levels is even essential for the understanding of the lower levels. Now you seem to turn things around. I would say that religion takes place on a rather complex level. But you seem to reduce the question whether there is a God or not to the question if there is a place for anybody who pushed the button of the big bang. That means that you go from the basic level (big bang) right to complex levels (religion).

*Answer*: One of the important contributions of modern physics is that it shows that we don't need a God for creative events like the origin of the Universe (a 'button-pusher') or for the origin of life. But there must be a deeper level of explanation, and if you like you can call this God, by which I refer to the nature of the laws of physics themselves and the fact that they could have been otherwise. To me the most persuasive evidence for the existence of something else is the fact that physics is not a closed and complete account of existence, which is conveyed to me in particularly persuasive manner in the fundamental laws of particle physics.

*Question*: I think you make a category mistake when you say that the laws of Nature have to exist eternally in order to make our cosmological models comprehensible. But saying that some mathematical law exists means something different from saying that people or matter exist.

*Answer*: I emphasized the curious inversion that has taken place in physics. In nineteenth-century physics the particles and physical stuff were considered as being real. The reality is now vested in the laws. The wave function prevents us from ever knowing in the classical, exact sense for example all the attributes of an electron at one moment. So the Hamiltonian or the Lagrangian becomes the central object of study. The laws have a Platonic status in modern physics.

*Question*: Don't you agree that the only existence of laws is in textbooks, that they are representations that should not be confused with what they represent.

*Answer*: I strongly disagree! You would have a hard job convincing a physicist that there are no real laws out there, only human projections. That is to reduce science to a charade. Physicists expose a *really existing* order in Nature, they do not merely impose an order *on* Nature. Of course, at any given time what we call laws – the things you read in the text books – are only approximate and tentative. They are, however, imperfect images of some really-existing things.

*Question* (Drees): You spoke about 'the dream that mankind would one day come to know the reason . . .'. Other dreams might be thought of, such as the dream that one day brothers and sisters will live well together, or that the lion and the lamb will play together. My question to you is what you think of such dreams in the context of cosmology. Is there a place in cosmology for dreams of goodness. Is there a notion of value somewhere in the Universe?

*Answer*: Some people attribute 'purpose' to our world. That would be an example of a kind of 'cosmological value'. I would call the Universe purposeful or meaningful, but never bored or impatient.

Can you make sense of a notion of 'goodness' in the Universe without turning to religion? You could, in a scientific sense, call our world the best of all possible worlds. Our world might be the most diverse and rich world, in a physical sense. You could try to quantify this in terms of organized complexity for instance.

*Question*: Are science and religion essentially different subjects?

*Answer*: The scientific world-view is clearly a product of the Western theological world-view, although scientists today rarely appreciate the theological origins of their assumptions. These include the existence of a lawlike order in Nature, the intelligibility and rationality of Nature, and the existence of a linear time dimension, starting with a definite origin and unfolding in a directed historical sequence. However, scientific methodology and the status of scientific evidence differ greatly from theology. An essential aspect of the scientific method is an inherent scepticism and provisionality. Scientific knowledge is always tentative, and scientists must always be prepared to change their minds in the light of new evidence or arguments. Scientific evidence must ultimately be both empirical and public. By contrast, theology is based on doctrine and received wisdom. To be secure in its foundations, theological doctrine should not change (though of course it does). Belief is more a matter of faith than evidence, and such evidence as there may be is often that of personal revelation or experience. Science and religion may find common subject matter (e.g. the origin of the Universe, the nature of human freedom) but they come at it from opposite poles.

*Remark* ('t Hooft): I would like to comment on Davies' amazement that the Universe is intelligible for us. I also used to be surprised about the fact that we humans have brains selected by evolution to be able to make the best bows and arrows so that we have a better chance at survival and procreate, and that those same brains can be used to study the cosmos and elementary particles. Now some time ago when I bought my first personal computer I first found in its manuals that it was especially designed to play computer games, and it was very efficient if you wanted to shoot aliens off the screen. But then, since it was a programmable computer, I was surprised how easy it was to reprogram it and use it to solve mathematical equations, work out cosmological models or properties of particles. So now I believe that this is a fundamental property of logic: it is universal. Once you can apply it in one problem you can apply it for everything.

**Questions to Mary Hesse**

*Question*: In the last few days, the question of realism has come up again and again. Bas van Fraassen has no need for realism. On the other side, Paul Feyerabend asserts that the scientist needs realism as a world-view in order to enable him to go on with his research. Without realism, he would drown in an ocean of anomalies. Why, according to you, is it so important to be a realist in both science and religion? Why is there a need for religious believers to have a transcendental reality?

*Answer*: In a way everyone is a realist, that is in the belief that we get hold of something true about the world when we construct regularities and find relationships that enable us to control small bits of the world. Of course this applies to everyday regularities that we must rely on, as well as to science. Fire rises and apples fall. But there is a different question, which is how far one believes that we have actually grasped the unique structure that lies behind the phenomena. I want to be agnostic here. I would not want to claim that science can describe much of it truly, and I think the same holds for models of God.

Positivists believe that what exists is exhausted by what we can have epistemological reasons for believing in, but this is too restrictive an ontology, if only because it neglects the open-ended and fragmentary nature of our experience. We have a sense for something beyond current experience and current types of reason, which motivates us to explore beyond them by means of hypotheses and models both in science and religion. This is a moderate realism. It does not follow that we can claim to have grasped the whole truth about what lies beyond, either globally or in precise local detail, and something like this is what strong realists have to claim if they speak about knowing *true* laws or theories.

As for why religious believers need a transcendental reality, I would draw a parallel between scientific and religious models. Science leads us to postulate the existence of material and structural realities beyond immediate phenomena. To have a religious belief, in the normal sense, just *is* to believe in something beyond the material as such, or at least beyond subjective and inter-personal experiences such as good and evil, altruism and duty, justice and mercy, wonder and love, which are the starting points of religion.

*Question*: Isn't realism needed to make science and religion intelligible?

*Answer*: No, in science I think sophisticated people may be able to operate in a formal and instrumental way with the phenomena, without transcendental reference. Perhaps mathematical cosmology is a good example of this, where theories and models can be taken as hypothetical games, to be judged by criteria of mathematical elegance and empirical fit, and not as potential descriptions of reality. There is also a social dimension to this. Social contructivists claim to make sense of science in terms solely of pragmatic social needs and ideological constraints, without real reference. Arguments for a purely social account are somewhat stronger in religion than in science, particularly because (Western) science produces a good deal of pragmatic consensus, whereas in religion there are many socially viable manifestations which are logically incompatible. Social anthropologists can make some sense of the diversity of religions as answering to social needs, without introducing notions of truth or transcendental reference. They are rather more successful in this type of explanation than are theologians who try to argue that one religion is nearer to the truth than others.

*Question*: You say you are a moderate realist in the sense that you are willing to say both 'yes' and 'no' to the question whether phlogiston exists. If you were to look at all of your experiences regarding the existence of a God (the religious experiences and their psychological embedding) you would be forced to say that probably there is no God.

*Answer*: The 'yes' or 'no' depends on the description of God. We can describe a substance as that which Priestley observed to be emitted when sulphuric acid was poured on zinc, and say that it exists, and that Priestley called it 'phlogiston'. Aquinas understood this interaction of experience and naming when he ended each of his five existence arguments 'and that is what all men call God'. But one might at the same time say 'no' to the existence of a God as described in some detailed religious creed, just as we would have to say 'no' to Priestley's detailed account of his phlogiston. Such an ambivalent attitude to statements of existence seems to be very common among both scientists and religious believers, and does not seem to me at all irrational.

*Question*: The concept of God as something that might be there but probably isn't, will make the concept incapable of doing what it is supposed to do in everyday life.

*Answer*: The concept 'probably' needs clarification here. I can perfectly well say that something like electrons exist with many of the properties we ascribe to them, but that probably the description of them given in any of our current theories is in detail false. Similarly I can say I don't believe in a God with some place in cosmological space–

time, or who is merely a stern lawgiver and judge. To bring out what descriptions of God are believable, one has to identify the needs and experiences people have which are answered by the religion they believe in, and then judge how far the religion makes sense relative to these needs and experiences. The crucial difference between the social or psychological constructivist's account of religion, and its transcendental interpretation, will then depend on whether one takes the needs, experiences, feelings of obligation, etc. to be features of human beings that transcend the material and the social, or not.

*Question* (van Fraassen): Every reasonable person is a realist in the sense that he believes that tables and persons exist. But you cannot reduce language to a sense level. Still it is important to look at our language and at what it presupposes. When I refer to a VHF-receiver, that is something highly theoretical. But when I buy one, I don't feel the need to tell the clerk that I'm agnostic about its existence. Suppose Lavoisier was in a crowded theatre after he had come to his oxygen theory and suddenly someone would yell 'phlogiston escaping!', you wouldn't expect him to sit down again, saying there is no such thing as phlogiston. I think there is no theory of language that makes it intelligible that we communicate successfully with radically defective languages. Does the idea of an analogical language help with that?

*Answer*: I entirely agree that language is crucial to all these problems. In natural languages, shifts of meaning are important and all-pervasive. There has been a philosophical obsession with terms with an essential meaning that stays the same in all contexts, and this is because univocity is necessary for logic, and philosophers have tended to regard logic as the crown of rationality. They believe we cannot deal with shifts of meaning unless these are explicitly distinguished. But Wittgenstein was right to point out that his 'family resemblance' terms can never be given necessary and sufficient conditions of correct application. Terms come to be used intelligibly as a result of learning how to modify and extend their meaning by analogies, metaphors and other figures of speech. We badly need a new theory of rhetoric, not just for the embellishments of literary language, but for descriptive language as such. This must go beyond Wittgenstein's 'language games' to uncover the intelligibility constraints upon metaphors that are pervasive in all language.

*Question*: If I understand you correctly, you distinguish between the merely metaphorical use of a metaphor like some kind of model, and on the other hand the metaphysical use of a metaphor, which would imply truth. What kind of realism are we left with then? You refer to yourself as a moderate realist, but metaphysical realism doesn't seem moderate and metaphorical realism doesn't seem like realism at all.

*Answer*: I believe that all descriptions are metaphorical, in the sense that *no* linguistic terms (except in logic and perhaps in some kinds of technical language) have necessary and sufficient conditions for correct application. Terms are always contextual, involving shifts of meaning with context, and requiring educated judgment about how to apply them in each context intelligibly and appropriately. So there is not the distinction you suggest between use of a metaphor in a model, whether scientific or religious, and its metaphysical use. In all cases the model or metaphysics may be more or less adequate, and in some cases we may be able to say that a metaphorical expression is (locally and approximately) *true*. We need a theory of truth that would enable us to say of van Fraassen's example, that 'phlogiston escaping' is (contextually) true, without going through the circumlocution 'what they *meant* to say was . . .' Such an interpretation of what they meant is unsatisfactory for a moderate realist, because (a) *they*, unlike Lavoisier, had not heard of oxygen and hydrogen, and (b) *we* ought not to claim that the circumlocution in our contemporary theory is set in concrete truth for ever. Unfortunately we are nowhere near having such a theory of 'truth', though in science we can describe the conditions it would have to satisfy, by giving criteria for theories that are good relative to current evidence and rational beliefs. With regard to metaphors of God, the criteria are less clear, because we don't know, or don't agree about, what 'good' models for God would look like.

*Question*: What are the phenomena you want our models of God to explain? I am not referring to religious behaviour here but to something like Paul Davies said about the surprising capacity of mathematics to describe the world.

*Answer*: Is the surprising capacity of mathematics to describe the world a *religious* problem? If it is, I would be quite content to 'deconstruct' it in the same way social theorists of religion do. This would involve pointing out how the history of mathematics, from measures of the pyramids onwards, has developed alongside physical applications, even in the case of the most abstract 'pure' mathematics. It would take a long time to argue this, but perhaps, briefly, one can regard pure mathematics as a large library of software for playing all kinds of games, some items of which can be selected and adapted to scientific use. There is no metaphysical mystery about this, only a manifestation of enormous human ingenuity.

We are able to describe relatively deep regularities and order in Nature, though not, I believe, with any guarantee that there is a rock-bottom (the sequence of theories may after all end in chaos). Insofar as there is mathematical order, it poses no religious problem over and above the presence of order that is apparent to the most superficial observation. All this can, indeed, be made an occasion of religious awe and wonder, but does not in itself justify any belief in a God who is relevant to the other problems and experiences normally understood as religious. To become a religious faith, wonder at the order of Nature requires belief in a God with other characteristics and for other reasons. The phenomena we chiefly want to explain with our models of God are not the same as those we want to explain by doing science. Other phenomena, like moral and social institutions, people's statements of belief, and people's internal spiritual needs, require explanation. We cannot find models of God by observing his immediate actions (he is a 'hidden God'), but only by interpreting such experiences as indicating some supra-natural meaning in the world and in the lives of human beings.

# Notes on contributors

**John D. Barrow** is Professor of Astronomy at the University of Sussex. His research interests are in cosmology, gravitation and astrophysics. His books include *The Anthropic Cosmological Principle* (Oxford, 1986), *The World Within the World* (Oxford, 1988), *Theories of Everything* (Oxford, 1991) and *Pi in the Sky* (Oxford, 1992).

**Paul Davies** is Professor of Natural Philosophy at the University of Adelaide. He previously held the Chair of Theoretical Physics at The University of Newcastle upon Tyne. His main research interests are in cosmology and the theory of gravitation. Davies is well known for his books on the conceptual foundations of physical science, the most recent of which is *The Mind of God* (Simon & Schuster, London & New York, 1992).

**Dennis Dieks** is Professor of the Foundations and Philosophy of Science – especially the Foundations and Philosophy of Physics – at Utrecht University. He is particularly interested in the foundations and philosophy of quantum mechanics and space–time theories. He has published numerous essays on these topics and other aspects of the philosophy of physics.

**Willem B. Drees** is on the academic staff of the Bezinningscentrum, a centre for the interdisciplinary study of science, society and religion, at the Vrije Universiteit in Amsterdam, the Netherlands. He holds degrees in theoretical physics and in theology. He is the author of *Beyond the Big Bang: Quantum Cosmologies and God* (Open Court, La Salle IL, 1990) and some books in Dutch. He is currently writing a study on naturalistic views of reality, including religion.

**Paul Feyerabend** was Professor of Philosophy at the University of California, Berkeley, and Professor of Philosophy of Science at the

Eidgenössische Technische Hochschule, Zürich. His books include *Against Method, Science in a Free Society, Wissenschaft als Kunst, Farewell to Reason,* and *Three Dialogues on Knowledge*. He had just finished his autobiography: *Killing Time*, when he died on 11 February 1994.

**Bas C. van Fraassen** is Professor of Philosophy at Princeton University, and works mainly in philosophy of science but has also written on literature. His most recent books are *The Scientific Image* (1980), *Laws and Symmetry* (1989), and *Quantum Mechanics: An Empiricist View* (1991).

**Mary B. Hesse** is Emeritus Professor of Philosophy of Science, University of Cambridge. Her main interests are history and philosophy of physics, philosophy of social sciences, and problems of scientific inference. Her books include *Models and Analogies in Science, The Structure of Scientific Inference,* and *Revolutions and Reconstructions in Philosophy of Science*. She is a Fellow of the British Academy and a Member of the Academia Europaea.

**Gerard 't Hooft** is Professor of Theoretical Physics at Utrecht University in the Netherlands. He first became known from his work on renormalizable gauge theories for elementary particles in 1971. Later he turned his attention more and more towards understanding the gravitational force. He has received several scientific prizes and an honorary degree.

**Ernan McMullin** is Director of the Reilly Center for Science, Technology, and Values, and Professor of Philosophy at the University of Notre Dame. His research interests lie in the philosophy of science, and the relations of science and religious belief. His most recent book is *The Inference That Makes Science* (1992). He is a Fellow of the American Academy of Arts and Sciences.

# Bibliography

This bibliography contains a selection of books by, or recommended by, the contributors to this volume. Also included are the books that are mentioned in the Introduction or in the Discussion.

Michael A. Arbib and Mary B. Hesse, *The Construction of Reality* (Cambridge University Press, 1986)

John D. Barrow, *Theories of Everything* (Oxford University Press, 1991)

John D. Barrow, *Pi in the Sky* (Oxford University Press, 1992 and Penguin Books, 1993)

John S. Bell, *Speakable and Unspeakable in Quantum Mechanics* (Cambridge University Press, 1987)

Fritjof Capra, *The Tao of Physics* (Shambala, Boulder, 1975)

Paul M. Churchland, *Scientific Realism and the Plasticity of Mind* (Cambridge University Press, 1979)

Paul M. Churchland and Clifford A. Hooker (eds.), *Images of Science: essays on realism and empiricism, with a reply from Bas C. van Fraassen* (University of Chicago Press, 1985)

Paul Davies, *The Cosmic Blueprint* (Simon and Schuster, New York, 1988)

Paul Davies, *The Mind of God* (Simon and Schuster, New York, 1992)

Willem B. Drees, *Beyond the Big Bang: quantum cosmologies and God* (Open Court, La Salle IL, 1990)

Paul Feyerabend, *Against Method* (New Left Books, London, 1975 and Verso, London, 1978)

Paul Feyerabend, *Farewell to Reason* (Verso, London, 1987)

Paul Feyerabend, *Three Dialogues on Knowledge* (Basil Blackwell, Oxford, 1991)

Richard P. Feynman, *What Do You Care What Other People Think?* (Bantam Books, New York, 1989)

Bas C. van Fraassen, *The Scientific Image* (Oxford University Press, 1980)

Bas C. van Fraassen, *Laws and Symmetry* (Oxford University Press, 1989)

Stephen W. Hawking, *A Brief History of Time* (Bantam Books, London 1988)

Gerard 't Hooft, *The Building Blocks of Creation* (*Provisional title*), translated from the Dutch, to be published by Cambridge University Press.

Jarrett Leplin (ed.), *Scientific Realism* (University of California Press, Berkeley, 1984)
Ernan McMullin, *The Inference That Makes Science* (Marquette University Press, Milwaukee, 1992)
Hilary Putnam, *The Many Faces of Realism* (Open Court, La Salle IL, 1987)
Steven Weinberg, *Dreams of a Final Theory* (Hutchinson Radius, London, 1993)
John Ziman, *Reliable Knowledge* (Cambridge University Press, 1978)

# Index

# Dimensions

basic 3

$E = MC^2$ time is a 4th dimension

↖ energy = mass X speed light $^2$

superstring theory =

10 dimensions
- 4 traditional
- 6 too small to see

Perception - we can only see a 3rd dimension (ie. a bee flying) we can not experience it